THE RESTLESS ATOM

Born in Pleasantville, New York, in 1906, Dr. Alfred Romer was educated at Williams College (B.A., 1928) and California Institute of Technology (Ph.D., 1935). His first research on radioactivity came shortly after he received his B.A. and returned to Williams as an Assistant in Chemistry. Working with a problem on the behavior of lead in dilute solutions, he made up solutions, treated them, evaporated them, and measured radioactivity on the dry dishes, learning in the process about the curves of radioactive growth and decay. That experience, says Dr. Romer, is one source of the present work. Another source has been Dr. Romer's experience in teaching, at Whittier College and at St. Lawrence University, a special course in physics to non-science students. "I came very early to feel," says Dr. Romer, "that physics could not be made real to these students if we started with fundamental laws, but that something might come of treating physics as a natural science which grew out of experiments. The best experiments to present seemed to be those which actually happened, the research experiments of the pioneer investigators." Altogether Dr. Romer has been investigating radioactivity—in the laboratory, in the classroom, and in the library—for more than 20 years.

Dr. Romer has contributed to *Isis* and the *American Journal of Physics*. For the last ten years he has been teaching at St. Lawrence University in Canton, N.Y., where he resides with his wife and three children.

THE RESTLESS ATOM

Alfred Romer

Published by Anchor Books
Doubleday & Company, Inc.
Garden City, New York
1960

Available to secondary school
students and teachers through
Wesleyan University Press Incorporated
Columbus 16, Ohio

COVER DESIGN BY GEORGE GIUSTI
TYPOGRAPHY BY EDWARD GOREY
ILLUSTRATIONS BY R. PAUL LARKIN

Library of Congress Catalog Card Number 60–10681

*Copyright © 1960 by Educational
Services Incorporated*

THE SCIENCE STUDY SERIES

The Science Study Series offers to students and to the general public the writing of distinguished authors on the most stirring and fundamental topics of physics, from the smallest known particles to the whole universe. Some of the books tell of the role of physics in the world of man, his technology and civilization. Others are biographical in nature, telling the fascinating stories of the great discoverers and their discoveries. All the authors have been selected both for expertness in the fields they discuss and for ability to communicate their special knowledge and their own views in an interesting way. The primary purpose of these books is to provide a survey of physics within the grasp of the young student or the layman. Many of the books, it is hoped, will encourage the reader to make his own investigations of natural phenomena.

These books are published as part of a fresh approach to the teaching and study of physics. At the Massachusetts Institute of Technology during 1956 a group of physicists, high school teachers, journalists, apparatus designers, film producers, and other specialists organized the Physical Science Study Committee, now operating as a part of Educational Services Incorporated, Watertown, Massachusetts. They pooled their knowledge and experience toward the design and creation of aids to the learning of physics. Initially their effort was supported by the National Science Foundation, which has continued to aid the program. The Ford Foundation, the Fund for the Advancement of Education, and the Alfred P. Sloan Foundation have also given support. The Committee is creating a textbook, an extensive film series, a laboratory guide, especially designed apparatus, and a teacher's source book for a new integrated secondary school physics program which is undergoing continuous evaluation with secondary school teachers.

The Series is guided by the Board of Editors consisting of Paul F. Brandwein, the Conservation Foundation and Harcourt, Brace and Company; John H. Durston, Educational Services Incorporated; Francis L. Friedman, Massachusetts Institute of Technology; Samuel A. Goudsmit, Brookhaven National Laboratory; Bruce F. Kingsbury, Educational Services Incorporated; Philippe Le-Corbeiller, Harvard University; and Gerard Piel, *Scientific American.*

PREFACE

Everything that we know in physics we have learned from experiments, and not from a small handful of them either. Behind even a tiny scrap of knowledge there may lie dozens and dozens. Of them all, the ones we have learned the most from in the end are those that went wrong. They were not always very good experiments, it is not always clear what a man was up to when he began one, nor what he did while it was going on. The important thing is that he came out with a puzzle. Then he had to think, and try other experiments, and think again, to discover what had really happened. As his ideas began to sort themselves, he could see more clearly what he ought to try and how he ought to go about it, and if everything went well, he might wind up with a superb piece of work. His last experiment, that is to say, would be beautifully planned, admirably carried out, crystal-clear in its meaning, leading to exactly the results that had been predicted. It would be a masterpiece, and it would tell him surely what he really knew already, that his ideas were finally right, and that he understood the situation.

This is a book about the experiments by which we have gained one section of our knowledge of atoms and the way in which they behave. You will find all sorts of experiments in it, puzzling ones and enlightening ones, easy ones and hard ones, big ones, little ones, and even a few masterpieces. Weaving in and out among the experiments you will find the arguments by which they were interpreted, the chains of reasoning which tied them together, and the ideas which had to be invented to make sense out of apparent nonsense. This is a book about one part of physics as it actually grew.

There is one difficulty in telling the story of things as they happened, and it is related to the problem of the perfect map. There was a small country once that planned for itself a perfect map to show everything just as it was, on a scale of one foot to one foot. So they made the map, and when it was done, they found to their dismay that they had no place where they could spread it out and admire it.

Like a map, a story tries to bring together the representations of large things in a small space. Like a map, it tries to leave out some details to make the connections between others clearer. That has happened here. It has not seemed helpful to describe every experiment or to mention every man who moved things along, neither has it seemed helpful to speak of those who hindered with misleading experiments and mistaken notions.

Within these limits, however, this story is as true as it could be made. It has been pieced together from the proper raw material of modern physics, the research reports of the investigators themselves. Its author is very grateful then for the existence of libraries, where all this raw material stands neatly stacked, ready for any hand to take down and any eye to read.

He is grateful to people too, to many who have helped him, and to three in particular: to Professor John F. Smith of St. Lawrence University, who freely allowed him to try his ideas in the classroom; to Dr. Donald C. Peckham, who has continued to encourage him; and most of all to his wife, who has listened to him when he wanted to talk, typed for him when he had a manuscript, and always found the ways to keep him going.

CONTENTS

11

CONTENTS

CONTENTS

1. By Way of Getting Started

Early in the new year of 1896, all over the world, people opened their newspapers to read a little story from Vienna. The report said that a German professor named Routgen had discovered a way of photographing hidden things, even to the bones within a living, human hand. It was a startling story, especially since it happened to be true. In a very few weeks laboratories in every country began to turn out pictures of bones: bones of hands and bones of feet, bones of arms and of legs and of anything else that could be managed in the human anatomy. Surgeons saw the usefulness of this strange photography, and (once the spelling of his name had been corrected) Professor Wilhelm Conrad Röntgen of the University of Würzburg became one of the most celebrated men of the day.

Our business in this book is with atomic physics, or at any rate with a part of it. It is to be about atoms which changed their nature, which in the ancient language of the alchemists transmuted themselves from being atoms of one element into atoms of another. It may seem odd then to begin it with a piece of medical history, but there is a good reason why this is necessary.

You cannot develop a science about something you do not believe in. In 1896 not many physicists believed in atoms, and no one at all believed in transmutations. There had been a time when transmutations had seemed reasonable, when it looked as though only a little change in color would be needed to convert a heavy, dull metal like lead into a heavy, bright metal like gold. The world is full of more spectacular changes. Lead can be roasted into a crumbling, yellow-red powder called litharge; gold can be dissolved in the proper mixture of acids. Yet out of litharge, you can get only lead back and from the acids, only gold. Somehow, the transmutations of the alchemists were never quite reliable, and the dream of altering substances in this style had to be abandoned.

The new scientists (who preferred to be called chemists) came instead to look on gold and lead and iron and sulfur as unchangeable elements, the basic substances out of which the many materials that fill our world were assembled. As knowledge of the elements began to grow, the ancient notion of atoms was put to use again, turning out to be so valuable that by 1896 any chemist could give you chapters of information on the way in which atoms behaved.

Each element (he would say) represented a single variety of atoms, and these atoms could combine in special patterns to produce the molecules of more complicated substances. Two atoms of hydrogen bound to one of oxygen, for example, made the molecule of water; three of oxygen to

16

two of iron the molecule of iron-rust; twelve of carbon, twenty-two of hydrogen, and eleven of oxygen the molecule of sugar.

What made the atoms of one element different from those of another was their chemical behavior, that is, their manner of combining with the atoms of other elements, and the sorts of substances those combinations produced. Beyond that, for each element, the atoms had a special weight of their own.

This is not to say that you could weigh a single atom. Atoms were far too small to be handled one by one. But there were experiments in which you weighed the total amounts of different elements that combined together. When you burned hydrogen to make water, 2 grams of hydrogen (or 2 ounces, if you prefer) would unite with 16 grams (or 16 ounces) of oxygen to make 18 grams (or 18 ounces) of water. Since the water molecule held two hydrogen atoms to each one of oxygen, this seemed to say that the oxygen atom was 16 times as heavy as the atom of hydrogen. When you heated copper red-hot you would need 63.6 grams of it to take up 16 grams of oxygen and make 79.6 grams of pure copper oxide. Again, with 63.6 grams of copper and 32.1 grams of sulfur, you could make 95.7 grams of copper sulfide.

From these and hundreds of other experiments it was possible to work out a series of numbers telling not the actual weight of any atom, but the relative heaviness of one compared with another. If you gave hydrogen the number 1, then (as you have just seen) oxygen would be assigned 16, sulfur 32.1, and copper 63.6. These numbers were the atomic weights, and each element, as it turned out, had a special atomic weight of its own.

I	II	III	IV	V	VI	VII	VIII
Li 7.03							
	Be 9.08	B 10.95	C 12.0	N 14.04	O 16.0	F 19.06	
Na 23.05	Mg 24.3	Al 27.1	Si 28.4	P 31.02	S 32.1	Cl 35.45	
K 39.11	Ca 40.1	Sc 44.1	Ti 48.2	V 51.4	Cr 52.14	Mn 55.0	Fe 56.0 Co 58.9 Ni 58.1
Cu 63.6	Zn 65.4	Ga 69.9	Ge 72.5	As 75.0	Se 79.0	Br 80.0	
Rb 85.4	Sr 87.6	Y 89.0	Zr 90.4	Nb 93.7	Mo 96.0	?	Ru 101.7 Rh 103.0 Pd 106.4
Ag 107.9	Cd 112.0	In 113.9	Sn 119.1	Sb 120.4	Te 127.5	I 126.9	
Cs 132.9	Ba 137.4	La 138.6 Yb 173.2	Ce 140.2	—	—	—	
—	—	—	—	Ta 182.8	W 184.5	—	Os 191 Ir 193.1 Pt 194.9
Au 197.2	Hg 200.0	Tl 204.1	Pb 206.9	Bi 208.1	—	—	
—	—	—	Th 232.6	—	U 239.6		

This was enough to make the atomic weight important, but beyond that, atomic weights were involved in a very curious discovery made in 1869 by Dmitri Mendeleyev of the University of St. Petersburg (now Leningrad) in Russia. He made a list of all the elements in the order of their atomic weights, from the lowest to the highest, and as he entered them in the table, he found that every so often the same kind of chemical behavior would come up. Hydrogen had to be set aside in a class by itself, and the list proper began with lithium, a metal whose oxide dissolved in water to make a strong alkali. Seven elements farther down the list came sodium, another alkali-producing metal, and seven elements beyond that came potassium, a third. The next element after lithium was beryllium, the next after sodium was magnesium, the next after potassium, calcium, and these three again made a group whose chemical behavior was very much alike. In fact, if Mendeleyev chopped his list up into sets of seven, and stacked these sets in parallel rows, one below the other, each vertical

Fig. 1. THE PERIODIC TABLE. *Dimitri Mendeleyev (1834–1907), a great Russian chemist, discovered in 1869 that the then known elements could be arranged by order of atomic weights in a table that revealed an unexpected symmetry of chemical behavior. This is the Periodic Table in its 1898 version. Vacant spaces in the table led to the hypothesis that unknown elements of the appropriate characteristics must exist, and the search for those unknowns had a prominent part in the discovery of radioactivity. In the table the names of the elements are concealed behind the chemist's conventional symbols.*

19

column contained a group, like lithium-sodium-potassium, whose members had very much the same sort of chemical behavior. With heavier elements, complications came in; it was necessary to alternate rows of ten with rows of seven, but the columns still filled with elements that obviously belonged together. (Fig. 1)

To speak precisely, this is not quite correct. Here and there, Mendeleyev had to force the fit. After calcium the next element on his list was titanium, and chemically speaking titanium belonged below carbon and silicon, so he pushed it into place by inventing a new element which he called "eka-boron" to lie below boron and aluminum. In the next row down, when he had fitted zinc into the beryllium-magnesium-calcium column, he came to arsenic, which plainly belonged under nitrogen, phosphorus, and vanadium. Two more invented elements, "eka-aluminum" and "eka-silicon," were needed to fill the space between.

This was bold enough, but Mendeleyev went further to describe these invented elements, telling what their atomic weights would be and into what chemical reactions they would enter. Then, as the years went by, one after another of these purely theoretical elements turned up, each answering neatly to Mendeleyev's description. François Lecoq de Boisbaudran found eka-aluminum in 1875 and named it gallium. In 1879, Lars Fredrik Nilson came upon eka-boron and named it scandium. Finally, in 1886, Clemens Winkler ran down eka-silicon and named it germanium.

Mendeleyev's arrangement, his Periodic Table of the elements, was something to be taken very seriously then. Unless a substance could be fitted into its scheme, unless its atomic weight and chemical behavior agreed when it was entered in the proper row and column, it could hardly be accepted as a respectable element.

Since the chemists knew all this about atoms, why were the physicists so skeptical? They were at home with molecules; they could imagine them stiffly linked together to form solids, sliding easily past one another in liquids, or wandering independently through empty space to make up gases. They could imagine these things, and from their imagining pass on to invent experiments which tested how well their ideas agreed with the actual behavior of molecules. With atoms, however, no idea seemed to work. There was no way to guess what sort of forces bound them into molecules, no way to guess why oxygen combined with iron but not with gold. It was impossible to imagine why the weight of an atom should be so important, why different weights gave their atoms different chemical behavior, and why in the rows and columns of the Periodic Table the same chemical behavior came around and around again.

In spite of all the chemist knew, the physicist found very little about the atom that he could fix his mind on. There seemed no way to bring its doings under the ordinary laws of physics, nor any way to invent new laws for it. It seemed necessary to leave it out of physics altogether, and what gets left out can hardly seem real.

Very well then, atomic physics could not exist, and transmutations were impossible. How, out of that state of the scientific mind, could a physics of transmuting atoms get started? The answer is that it only happened, that it grew by itself out of pure accident and curiosity. Somebody noticed something odd; someone else, growing curious, investigated. For no particular reason, inexplicable facts piled slowly up, until one day it appeared that the notion of transmuting atoms would make reasonable sense of everything.

This makes the story a little hard to tell. It be-

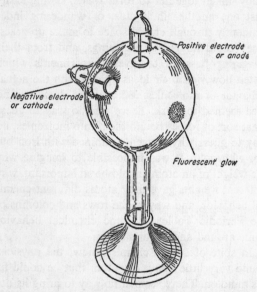

Fig. 2. SIMPLE CATHODE-RAY TUBE. *The tube is an evacuated glass bulb with a negative electrode, or "cathode," and a positive electrode, or "anode," sealed through the glass wall. The location of the anode is not important. When a fairly high voltage is applied, the cathode rays, which are streams of electrons, come off the cathode at right angles to its surface. Where they strike the glass wall a fluorescent glow is excited.*

gins nowhere in particular, it wanders off in unexpected directions, it comes out in surprising places, and never seems to arrive at anything settled. Nevertheless, if you will stay with it, taking the jolts and swerves as they come, you will discover in the end that you have been following the right road all the time.

So let us return to Röntgen. How he first came upon his rays we do not know, but it was probably by accident when he was busy about something else. The apparatus he needed to produce them was common and likely to be found in any university laboratory. He used a spark coil to supply electricity at high voltage, a cathode-ray tube to discharge it through, and that was all. The "tube" was simply a bulb of glass, which might be round or sausage-shaped or pear-shaped, pumped down to a good vacuum and provided with a pair of metal "electrodes" for the electric discharge to pass between. (Fig. 2)

It was at one of these electrodes, the "cathode," where negative electricity jumped off to the scanty gas remaining in the tube, that the cathode rays came into being, and they stretched away at right angles to its surface. If the walls of the tube were close enough for the cathode rays to reach them, then under the play of those rays the glass lit up with a fluorescent glow, which was green for tubes made of English lime glass and blue for the lead glass of the Germans. Here, in this fluorescent glow, Röntgen's X-rays were produced.

However Röntgen first happened to notice them, the important thing he did was to investigate. He found where they were produced; he found how they traveled in straight lines, how they could excite a fluorescent glow in a particular compound, called barium platinocyanide; how they would penetrate some materials and be stopped by others, so that familiar objects cast very strange shadows across the glow they excited. He found that they would expose a photographic plate (film had been invented then, but everyone preferred glass plates for serious work), and he photographed the shadows, making strange, new pictures of the insides of things. During the last two months of 1895 he worked at top speed, and by Christmas he felt ready to make an announcement.

It is understood in science that the first man to make his discovery public may claim the credit for it. He may claim no more than he announces, however, and once he has announced it, it stands over his name, right or wrong, forever. Röntgen now was sure of what he knew, and he chose the quickest of all possible ways of getting it into print. There was in Würzburg a scientific society which met for the reading of "papers" (as research reports are usually called) and published them later in its *Proceedings*. On the Saturday after Christmas, Röntgen called on the secretary of this society, who accepted his paper and sent it to the printer, to be set up in type and run off at once as a ten-page pamphlet. On New Year's Day, Röntgen mailed copies of this pamphlet to the leading physicists of Europe, and into each envelope he slipped a handful of the pictures he had taken, the first X-ray pictures in the world. It was from the pamphlet sent to Vienna that word reached the newspapers, and this is how it happened that this German discovery was first made known in Austria.

2. The Penetrating Rays of Henri Becquerel

Our business, however, is with the copy of Röntgen's pamphlet that went to Paris, to the mathematical physicist Henri Poincaré. In Paris was the *Académie des Sciences,* whose seventy-eight members were the most distinguished scientists of France, and which stood at the center of all French science. It met on Mondays for the reading of papers (which it published within two weeks), and there, on the afternoon of January 20, 1896, the Academicians had the pleasure of seeing the first French X-ray of the bones of a hand, the work of two physicians named Oudin and Barthélemy. The pictures led to talk and the talk to questions, which Poincaré, of course, could answer.

Among the curious listeners was Henri Becquerel, an Academician as his father and grandfather had been and, like his father and grandfather also, Professor of Physics at the Museum of Natural History. What interested him was the report that the X-rays arose in the fluorescent spot on the wall of the cathode-ray tube. The fluorescence produced by light was one of the effects his father had investigated, and he himself had worked with it a little. If the fluorescence of the cathode rays contained X-rays, then X-rays might be produced in other varieties of the fluorescent glow.

So Röntgen's publication accomplished its work. To a total stranger it had given a new idea, and Becquerel went back to the Museum to put his idea to the test.

For a month he found nothing, and then for a new set of experiments he happened to choose as his fluorescent material some crystals of potassium uranyl sulfate. This is a complicated compound of potassium, uranium, oxygen, and sulfur, whose crystals (as he knew from personal experience) would glow under ultraviolet light. To detect the penetrating rays he still hoped for, he took a photographic plate, wrapping it in heavy black paper to screen it from ordinary light. For the ultraviolet light to excite the fluorescence of his crystals he chose sunshine, and he set the plates outside his window with the crystals lying above the paper wrappings. Hours later he took them in, and as he developed them under the red light of his darkroom, he was pleased to see the grayish smudges which slowly grew on their creamy surfaces wherever a crystal had lain.

He tried again, laying a coin or a bit of metal pierced with holes below each crystal, and now he saw those metal objects silhouetted in light patches on the darker gray around them. In a third trial, he set each crystal on a thin slip of glass to act as a barrier against any vapors which the sun's heat might have driven through the pores of the paper to blacken the plate by chemical action. Once more the plates darkened as though the glass were not there, and Becquerel was confident that he had found a penetrating ray which was produced by light. On February 24, at the next session of the *Académie* he reported it.

Notice how neatly it all worked out. Becquerel had made an hypothesis that X-rays were a normal part of fluorescence. The hypothesis had suggested an experiment, and the experiment had given exactly the results he predicted. It was as pretty and as misleading a piece of scientific work as you could ask. Luckily, Becquerel went on with new experiments, and even before his announcement appeared in print, he had learned a good deal more about his rays and was a good deal more perplexed.

In the next three days the weather changed. Wednesday's plates were hardly ready when clouds came over the sun, and into a drawer went plates, black paper, crystals and all. There they lay in the dark until Sunday, and in the dark, as Becquerel knew, nothing could happen. Potassium uranyl sulfate would glow only while the ultraviolet light fell on it; when that light was shut off, the fluorescent glow ceased within a hundredth of a second. Even so, when Sunday came, Becquerel, with a kind of methodical impatience, pulled out the unused plates and developed them anyway. What leaped up before his eyes were patches far blacker than he yet had seen. Even without light the crystals seemed able to send out their rays, and when he ran through the experiments once more in the total blackness of his darkroom, he found that this was true.

It was true and inexplicable, and all he could do was investigate. Some of his crystals he laid away in darkness to see how long it might take their penetrating rays to fade. Whenever he tested them, in the hours and days and weeks that followed, there were always rays pouring out vigorously. He tried other fluorescent materials, and whenever they contained uranium, he found his rays, but when they were made with calcium or zinc he did not. He tried uranium compounds that were not fluorescent, and from them, oddly enough, the rays appeared again.

What was puzzling about all this was the energy involved. It took energy to expose the photographic plates, energy which the crystals had somehow stored away. Becquerel would have liked to know how that energy entered a crystal, what he needed to do to start a crystal going, but none that he had seemed ready to run down. Shut away in his darkroom, he tried the trick of gently heating a crystal of uranyl nitrate until the water molecules which were built into its structure were set free by the warmth, and the crystal dissolved at last in its own "water of crystallization." That might have been expected to set free any stored-up energy, but when the test tube cooled and the uranyl nitrate recrystallized in the darkness, it regained its power to give out the rays. It was truer in fact to say that it kept it, for he presently found that rays came from the solution as freely as from the solid crystals.

The one constant thing in all his experiments was the presence of uranium. So long as his material contained uranium it did not matter whether it was fluorescent or not, whether it lay in light or darkness, whether it was solid or in solution. It seemed to Becquerel worth trying whether pure, metallic uranium might not give the rays also. Pure uranium did not exist, but, as it happened, Henri Moissan of the School of Pharmacy in Paris was busy at the moment on a new process for refining it. Becquerel waited, and when Moissan succeeded in early May, he tried out a disc of simple, uncombined, uranium metal. Its rays were more intense than any he had ever seen.

It was true, and yet altogether odd (as he pointed out) that a pure metal should have the power to give out rays from some unknown source of stored-up energy.

3. The Curies and Their Two New Elements

Since they did not give pictures of bones, Becquerel's rays were not nearly as fascinating as Röntgen's, and no one else saw any profit in studying the mysterious penetrating rays from uranium. Perhaps this is why they attracted Marie Curie near the end of 1897 when they had lain neglected for nearly a year and a half.

Madame Curie had begun life as Marya Sklodowska, the daughter of a teacher of mathematics and physics in Warsaw, in what was then Russian Poland. A driving ambition to study those subjects had sent her to Paris, where she had taken her preliminary degrees at the Sorbonne, and where she had met and married Pierre Curie, Professor of Physics at a technical school which the city of Paris maintained, the Municipal School for Industrial Physics and Chemistry. Now, after the birth of her first daughter, she was anxious to go on. The next stage would be the doctor's degree, and in France that required a long and elaborate piece of private research. A topic in which no one else was interested would give her an ideal project, since there was little danger that an unknown competitor in some other laboratory might solve its problems first.

She did not plan to work by photography as Becquerel had done, but to detect the rays by another property he had discovered. This was their curious ability to discharge electrified bodies. It was as though they managed to convert the air through which they passed from an insulator to a conductor of electricity, and in this effect she saw the possibility also of gauging their intensity. To do it, she would have to measure exceedingly delicate currents, but she had an excellent instrument for just such work, an improved electrometer, which Pierre Curie and his brother Jacques had designed. (Fig. 3)

She began with Moissan's uranium metal, and tried, like Becquerel, to find the source of its energy, but no heating of the disc, nor any exposure to light or X-rays managed to change the strength of its rays. At last, in February, she turned to something new. Becquerel had shown that the rays came from uranium in whatever state he had it. It occurred to her that the ray-giving might be a power other metals shared, and she began a hunt which went on and on with no particular success. Sometimes she tested pure metals, sometimes minerals just as they came from the mine, sometimes the carefully purified compounds of the chemical manufacturer. Again and again she found nothing, with the one very odd exception of pitchblende.

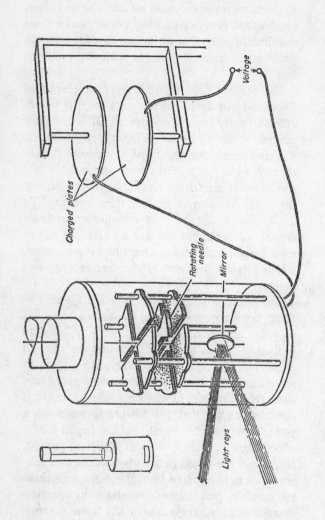

Charged plates

Voltage

Rotating needle

Mirror

Light rays

Pitchblende is an ore of uranium, but since its uranium had to share space inside the crystals with oxygen, not to mention a number of assorted impurities, it was most astonishing to find it giving out rays with considerably more intensity than the pure metal. She ran a long series of measurements with purified compounds of uranium and found only what might have been expected, that the more other elements diluted the uranium, the weaker were the rays. Yet quite regularly uranium minerals gave very strong rays. Just possibly the com-

---

Fig. 3. PIERRE CURIE'S ELECTROMETER. *With this apparatus Marie Curie measured the intensity of rays given off by uranium ores and pitchblende. The device at the right, a form of ion-collector, consisted of two insulated plates, on the lower one of which was spread the ray-giving material. The meter proper, center and left, had eight quarter-circle metal plates arranged on insulating rods to form four quadrants. Between the upper and lower sets of quarter-circle plates swung a thin aluminum needle, suspended on a fine wire in a glass chimney and free to rotate. At the end of the wire below the needle was a small mirror, which rotated as the needle swung. Light reflected from the mirror showed the extent of the rotation. A high-voltage battery charged the lower insulated plate in the device at the right. Rays from the material on the lower insulated plate permitted flow of current to the upper plate, which was connected to one pair of quadrants in the meter. The aluminum needle was charged through the wire from another source, and as it was attracted or repelled by the charged quadrants, it swung sideways. The amount it swung was measured by movement of light reflected from the mirror.*

plex crystal structure of a mineral might somehow strengthen the ray-giving, but it was a lame argument, which she proceeded to demolish by cooking up an artificial chalcolite (copper uranyl phosphate) out of laboratory-pure reagents. The uranium it contained gave no stronger rays in the crystal form of the mineral than in the storage bottles on the shelves, and the imitation remained weaker than natural chalcolite.

This raised the possibility that an impurity in the minerals was contributing the extra rays, but when she searched the Periodic Table from end to end, she was able to find only two ray-giving elements, uranium and thorium—and there was no thorium to speak of in the minerals she was testing. Could the impurity be an undiscovered element?

Although Marie Curie did not spell out the argument in the paper reporting all this, which Professor Gabriel Lippmann of the Sorbonne read for her at the *Académie,* it was plain enough to anyone who looked at the Periodic Table. Gallium, scandium, and germanium had by no means filled up all the gaps. There was a particularly impressive set between bismuth near the bottom and uranium and thorium at the very end. Those two heavy metals, with the fantastic names of forgotten gods, were the only ray-givers of all the elements known. If an unknown, ray-giving element also existed, it might well fit one of the gaps in their neighborhood.

Tracking down a new element was a job for a chemist, for an experienced chemist who knew intimately all the varieties of behavior of all the known elements. Neither Marie nor Pierre Curie was a chemist at all, and Gustave Bémont, the man they turned to for advice, was only the laboratory instructor at the Municipal School. Yet the element they were looking for would be easy to find. It must be a ray-giver, and if it differed only from uranium and thorium, then it had to be new.

Pierre Curie had gradually been drawn into the electrometer measurements and all the puzzles they raised, and now he took his place as an equal partner in the hunt for the new element. Together the two Curies ground up some pitchblende, dissolved it in acid, and set to work. What they had to do was to sort out all the different elements that might be in the mineral, and the great device for this kind of sorting was the filter. Whenever they could produce a slushy mixture of a liquid with the undissolved grains of some stuff, they would pour this into a cone of paper set down in a glass funnel. Then the liquid would ooze through the pores of the paper and drip away while the undissolved grains were caught and held back. That gave them two different things in two different places, and they had begun to sort out the mixture.

When, for example, they bubbled the unpleasant but very useful gas called hydrogen sulfide into their solution, this reacted with a few, perhaps five, metals to form insoluble sulfides, and these "precipitated" out in a slimy mass, which could be filtered off. Here there would be neither uranium nor thorium, but still there was an activity of ray-giving—a "radioactivity" they were beginning to call it—on the filter paper, and they knew that they were on the track. After that it was a matter of routine, to re-dissolve, re-precipitate, and re-filter until the five different metals had been got into five different dishes. In this way they discovered that the radioactive substance went along with the bismuth.

It could not be bismuth, for they knew (and made sure again) that bismuth was not radioactive. Then, after a little experimenting, they found a way to coax the bismuth and the radioactive substance apart. If they formed them into sulfides again, sealed up the mixture in a vacuum in a hard glass tube, and heated it strongly, the radioactive material would evaporate; it left the bismuth sulfide behind and condensed in a dark stain at the cooler end of the tube.

This was little enough to go on, but it seemed clear that the radioactive substance was not uranium nor thorium nor bismuth. It might be proved an element yet, and in their report (which this time Becquerel read for them) they proposed to call it polonium.

Perhaps it was a foolish sentiment that prompted them to name an element for a vanished nation. Long before, the Kingdom of Poland had been divided among Austria, Russia, and Prussia, and there seemed no prospect that the three powerful empires ruling its fragments might ever disintegrate to set them free. Yet even in Marie Curie's lifetime, this same sort of stubborn and romantic patriotism did manage to bring a Polish nation back into being.

The "bismuth" activity of polonium was not the only radioactivity the Curies found in their pitchblende, however. There was another, which sorted out with barium, and this one yielded to a chemical process of separation. The trick was to form chlorides out of the barium and the new element, to dissolve in water as much of the mixed chlorides as possible, and then to pour alcohol into the saturated solution. This forced out some of the dissolved material as a white precipitate, which could be filtered off, and the clear "filtrate," which had dripped down below, could be evaporated to recover the rest. When they compared these two portions, there was always more radioactivity in the precipitate than in the material which had remained in solution.

It was only a partial separation, but by doing it over and over, they were able gradually to crowd more and more of the active material into a smaller and smaller sample, until at last they had hardly a pinch of white powder, and this precious pinch, weight for weight, was nine hundred times more radioactive than uranium metal.

If they were to prove that this new substance was an element, they must get an atomic weight for it, and that meant accumulating enough of it to weigh. The day when they could do that was still in the future, but in the meanwhile they might get some hint by taking its spectrum.

This "spectrum" is a kind of characteristic, atomic light. If you evaporate a substance with a hot flame or an electric spark, and if the atoms of the substance have enough energy to set themselves glowing, then the light they give out is colored, and colored in a unique way. When you disperse the light with a prism, you do not see a continuous band of blending tints running all across from red to violet, but a pattern of sharp, narrow, brilliantly colored lines, separated by wide spaces of absolute darkness. For each different element the pattern is different, and although in a mixture of elements the patterns become entangled, with care and patience an expert can distinguish one from another.

In Paris there was such an expert, Eugène Demarçay, a chemist, from whom Marie Curie already had borrowed samples of some of the rarer elements to test for radioactivity. As their material had grown more concentrated, the Curies had kept taking it to Demarçay. In the spectrum of their last, most active specimen he found a single line in the ultraviolet range which did not belong to the pattern of the barium composing the bulk of the material, nor of the platinum of the wire by which he drew his sparks, nor indeed to the pattern of any known element.

On this evidence, on the basis of its radioactivity, of its partial separation from barium, and its single spectral line, the Curies announced their second new element at the very end of 1898, and for the great intensity of its rays they named it radium.

Then it was time to begin again. They could hardly afford to buy more pitchblende, but luckily they found a cheaper substitute. Through the Austrian government they got as a gift some hundreds of pounds of residues from the uranium refinery at Joachimsthal in Bohemia. (It is called Jáchymov now, and is in Czechoslovakia.) There was no uranium, of course, in that brownish powder mixed with pine needles, but then it was the radium that they wanted. Back they went to their chemistry, dissolving and precipitating and dissolving again. It was quite as well that they did not realize how the pounds of residues would grow to tons before they could scrape together a weighable amount of pure radium chloride.

4. Ernest Rutherford and Temporary Radioactivities

You have met uranium now in the hands of Becquerel, and polonium and radium with the Curies, but so far you have heard little about thorium, the fourth radioactive element. It was December 26, 1898, when Becquerel reported the Curies' discovery of radium to the *Académie des Sciences,* and just at this time thorium was being investigated, in quite a different quarter of the globe, by R. B. Owens, Professor of Electrical Engineering at McGill University in Montreal.

Owens would be called a Yankee at McGill (in spite of the fact that he had been born in Maryland), and he had arrived there only that fall from a job at the University of Nebraska. He was twenty-eight, and he promptly became friends with another newcomer and contemporary, Ernest Rutherford, a twenty-seven-year-old Professor of Physics. It was Rutherford who had suggested the study of thorium to him, and it was Rutherford's methods that he was following as he worked.

We shall have to introduce Rutherford with a little more formality. He came from New Zealand, where he had established himself as something of a prodigy. On his graduation from Nelson College (the equivalent of an American preparatory school) he won awards not just in mathematics, physics and chemistry but in Latin, French, English literature and history, too. His university record, at Canterbury College, was equally distinguished. He went from degree to degree, and the magnetic experiments he began in his fourth year brought a research scholarship at Cambridge University at the end of his fifth. In the fall of 1895, he arrived at Cambridge to settle in at its Cavendish Laboratory, and he already had begun to make a reputation when the X-ray excitement broke a few months later.

Early in February 1896, Professor J. J. Thomson, the Director of the Cavendish Laboratory, discovered that X-rays could turn the air through which they passed into a conductor of electricity. Since conducting gases were a specialty of his, he thought he saw how the X-rays might act. The investigations he started went well. In two months, to speed the work, he called in Rutherford to be his personal assistant, and by the time summer was out the two had a good general notion of what was going on. Then they divided forces to work out details.

As their ideas developed over the next year and a half, they came to picture the process like this. When a beam of X-rays passed through air (or any other gas) the rays were able now and then to rip from one of the molecules a very tiny particle charged with negative electricity. (When Thomson

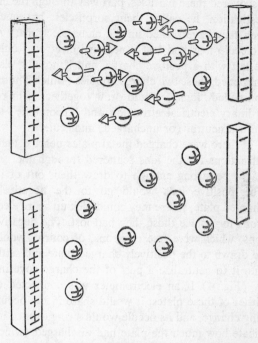

Fig. 4. COLLECTION OF IONS FROM AIR. *The upper diagram represents five positive ions plus moving toward the cathode at right, five negative ions (free electrons) minus moving toward the anode at left, and five neutral molecules plus or minus. (Actually there would be billions of neutral molecules for each pair of ions.) In the lower diagram the five free electrons have entered the anode and neutralized part of its charge. The five positive ions have taken electrons from the cathode and become neutral molecules. The end result makes it look as if five electrons had crossed the air space between the electrodes.*

discovered these particles, part way through the investigation, he called them corpuscles, but since then they have picked up the name electrons.) A molecule that had lost a negative electron would be left with a positive charge, and if nothing else happened, these two charged bodies (which Thomson called "ions") would draw together again by ordinary electrical attraction and "recombine" to form a neutral (or uncharged) molecule.

If there were charged metal plates nearby, their attractions for the ions scattered through the gas might be strong enough to draw them out of it. The positive ions would go to the negatively charged plate, where they could pick up spare electrons to replace those they had lost. The negative ions, which were free electrons, of course, would be drawn to the positively charged plate and settle into it to neutralize a part of the charge covering it. (Fig. 4) If an electrometer was connected to either of these plates, it would share in the occurring change, and its needle would swing over to indicate how much the plate had discharged (which is to say, how many ions it had picked up).

Rutherford's job with the collecting plates and electrometer had been to learn everything he could about the ions which the X-rays produced: how rapidly they came into being, how rapidly they recombined when left alone, and how swiftly they would move through the body of the gas under the pull of a charged collecting plate. This was a solid year's work, and when it was finished, he exchanged the X-ray tube for a dish of uranium oxide and spent another year on the ions produced by its rays. Between the two kinds of rays there were noticeable differences in behavior, which

Rutherford explored along the way, but the ions they produced were entirely the same.

During the summer of 1898, while the Curies were busy with their work on polonium, word had come to Cambridge of an opening at McGill. Rutherford (who was anxious to marry) had applied, and with Thomson's enthusiastic backing he was appointed. With all possible speed he finished his paper on the rays from uranium, and in September took a ship for his new position, where (as he wrote his fiancée in New Zealand, Miss Mary Newton) he was expected "to form a research school in order to knock the shine out of the Yankees!"

It was a challenge and an opportunity, but a student does not turn into a professor overnight. Four days later his mood was changing, and he added, "It sounds rather comic to myself to have to supervise the research of other men, but I hope I will get along all right."

Now he was doing it, and this was where Owens' investigation of thorium came in. The plan was that he would study the ionization its rays produced as Rutherford had done already with X-rays and the rays from uranium. The only real problem was a certain temperamental behavior which Rutherford had already noticed in the radioactivity of thorium oxide. Sometimes it seemed strong, sometimes it seemed weak, and it could change without warning in the middle of a measurement. By and by Owens ran down the trouble; the thorium oxide was sensitive to drafts. The only way to manage was to shut it up in an airtight box, wait a quarter of an hour for the last air current to die down, and only then begin to measure. If he blew a puff or two of air into the box, the ionizing power of the thorium oxide would drop abruptly, and it would take another quarter-hour before it steadied back to its original strength. In still air, however, the measurements were straightforward, and by spring Owens had finished the job. The rays from thorium produced exactly the same kind of ions, in very much the same way, as did the X-rays and the rays from uranium.

Then Owens sailed for England, to visit Thomson's laboratory, and Rutherford set about on his own to answer a question that had provoked his curiosity. What did an air puff do when it blew ionizing power away from the thorium oxide? He found the answer before winter settled in, but it was an odd one, involving two strange substances with temporary radioactivities. It was a very odd one, in fact, for although he could not see these substances, nor smell nor taste nor weigh them, he knew by circumstantial evidence that one was a gas and the other a solid.

To answer his question, he decided to hunt downwind, not investigating the thorium oxide itself, but rather the air that had blown past it. He arranged an ion-collector at one end of a long tube, with a paper package of thorium oxide at the other. The paper wrapping would keep any dusts of the oxide from joining the air and making complications, but the experiment still ought to work since Owens had discovered that paper was no screen against the air disturbances. (Fig. 5)

As might have been expected, no ions appeared in the collector until the air currents started to move, and not even then until the air had had time to work its way over from the package of thorium oxide. This was fair enough—it meant that the thorium oxide acted only on the air directly above it. What happened next was more interesting. When Rutherford shut the air streams off, leaving the ion-collector filled with air which had drifted past the paper package, it took about ten minutes for the ionization to die away. Since

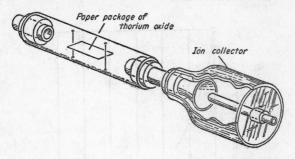

Fig. 5. RUTHERFORD'S "DOWNWIND" APPARATUS.
The air stream entered the opening at the left end of the device, picked up "emanation" from the packet of thorium oxide and carried it to the ion-collector. A 100-volt battery charged the metal walls of the collector, and ions carried the charge to the insulated rod at the center. The rod was connected to an electrometer.

only a few seconds would be needed to draw all the ions out of a gas, this meant that the supply of ions in the collector had been steadily renewed. Then something in the collector had been giving out ionizing rays, and this seemed to mean that a radioactive material had been carried along from the thorium oxide package.

What that material might be was the next problem. It could work its way out through the pores of the paper wrapping; it could work up through a thick layer of powdered thorium oxide. It was not trapped in the fibers of a cotton plug, nor washed away when the air stream was bubbled through water or sulfuric acid. It seemed finer than a dust, and Rutherford thought it might be a gas or vapor, but to be safely on the side of vagueness, he named it an "emanation."

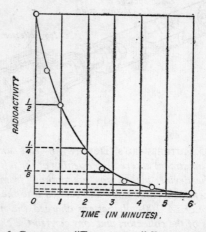

Fig. 6. GRAPH OF "EMANATION" RADIOACTIVITY. *The curve, an ideal one, shows how the radioactivity of the thorium oxide "emanation" died down with the passage of time. The small circles represent Rutherford's measurements. The height of the curve drops by half in every minute.*

In the face of a mystery there is nothing to do but investigate. The emanation was radioactive, and it lost its radioactivity after a time. Rutherford set out to discover how. Measuring the ionization over and over in a body of still air, he found that the radioactivity died down by a geometric progression in time, falling away by half with each minute that went by. (Fig. 6)

This is an interesting kind of behavior. When half of what you have vanishes in a minute, it means that you will lose much when you have

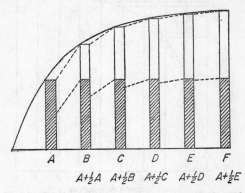

A   B   C   D   E   F

$A+\frac{1}{2}A$ $A+\frac{1}{2}B$ $A+\frac{1}{2}C$ $A+\frac{1}{2}D$ $A+\frac{1}{2}E$

Fig. 7. GRAPH OF "EMANATION" BUILD-UP. *The block A represents the new radioactivity brought in by the "emanation" released each minute from the thorium oxide. The total radioactivity, represented by blocks B, C, D, E, and F, is this amount plus half the radioactivity present the minute before. Thus, block F includes an A amount of fresh radioactivity plus half the E total of radioactivity existing a minute earlier.*

much, but when you have little, you will lose only a little. Putting this into mathematics, Rutherford saw an experiment with which he might test his understanding. Suppose he put some thorium oxide, tightly wrapped in paper, inside a closed ion-collector. The emanation would diffuse slowly through the paper, and the ion-collector would fill with it steadily. Once the emanation was out, it would lose radioactivity, but at the beginning, with little emanation in the ion-collector, the loss would be small. The gain in fresh emanation steadily diffusing from the package would more than make it

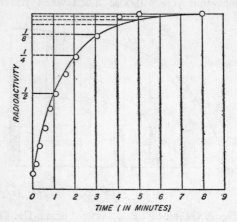

Fig. 8. GROWTH OF "EMANATION" RADIOACTIVITY *in a closed space. The circles represent Rutherford's actual measurements of the amount of radioactivity diffusing from a packet of thorium oxide in a closed ion-collector. The idealized curve shows that it made up half the difference between the existing value and the final value in each succeeding minute.*

up. So the amount of radioactivity would grow. As it grew, however, each minute's loss of half the activity would become a larger and larger quantity. (Fig. 7)

Doing the mathematics, Rutherford found that the radioactivity would grow toward this final value in a very particular way. If at any instant you saw how much the radioactivity lacked of its goal, you would know that one minute later it had made up half the lack, and in another minute, half of the remaining half. Once he knew how the experiment should work, Rutherford tried it out, and found that this was exactly the way it did. (Fig. 8)

Long before this, at about the time that he had begun to feel at home with the emanation, he had run into a complication. His ion-collector developed an electrical leak, but the new insulating plug, which should have cured it, was no help at all. The trouble turned out to be a good deal stranger; in some way the metal parts of the ion-collector had become radioactive.

Nothing like this had happened with X-rays or uranium; the power to "excite" radioactivity in nearby bodies seemed to belong only to thorium. Since the excited radioactivity appeared downwind and around corners, it was easy to argue that the direct rays of the thorium oxide did not produce it, and new experiments made this certain. It could appear when the thorium oxide was heavily screened with a deep pile of paper sheets, and this suggested a link with the emanation. In fact, thorium oxide could excite radioactivity freely only when it gave plenty of emanation. If the oxide was brought to a white heat, it lost its emanating power, and in the same process it lost its power to excite radioactivity.

Another strange thing was the way in which electrical forces moved the excited radioactivity. Negative charges seemed to attract it, and attract it rather strongly, so much so that it could be concentrated entirely on one tiny loop of negatively charged wire.

This suggested an interesting experiment. Rutherford made a long wooden box, down which an air stream could carry a slow current of emanation, a current so slow that the emanation lost a good share of its radioactivity in the time it took to go down the box. Along the bottom of the box he put a positively charged plate to drive the excited radioactivity away, and along the top he set four separate plates, one behind the other, all charged negatively so that the excited radioactivity would concentrate entirely on them. What appeared on any one of the plates, then, would be all that the emanation could excite as it drifted by that point. What Rutherford found when he tested the plates at the end of the experiment was a good deal of radioactivity on the first, and a shading down to rather little on the last. In fact, the excited radioactivity that the emanation produced on each plate was about proportional to the amount of radioactivity the emanation still kept to itself as it passed by.

He tried laying sheets of metal foil over the top of a plate that showed the excited radioactivity, and he found that the rays from it were a little more penetrating than those he had seen with uranium, or Owens with thorium. They seemed to come just from the surface of the plate, since a little scrubbing with sandpaper always made them weaker.

Like the radioactivity of the emanation, the new excited radioactivity was not permanent, but died away by a geometric progression in time. The only difference was in its speed; it needed eleven hours instead of one minute to accomplish the drop to half-value. When Rutherford exposed a negatively charged plate to the emanation from a dish of thorium oxide, and drew it out now and then for testing, the excited radioactivity on it grew toward a high, steady value, making up half of its goal in eleven hours. (Fig. 9)

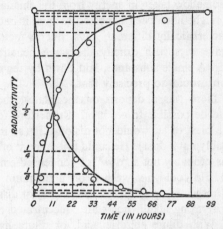

Fig. 9. GROWTH AND DECAY OF EXCITED RADIOACTIVITY. *For the rising or growth curve the time is the total exposure of Rutherford's collecting rod in the thorium oxide "emanation." For the falling or decay curve it is the time that has elapsed since the rod was removed from the "emanation." The circles represent actual measurements; the curves are idealized.*

Although he called it an "excited radioactivity," Rutherford was not at all sure that it was really excited by the emanation. There was a good deal to suggest that the rays came from some substance that the emanation laid down on the surfaces that it touched, but there was never anything that he could see under a microscope, nor any change in weight he could detect with a sensitive balance.

In one experiment he "excited" a good many different substances: copper, lead, platinum, aluminum, tin, brass, cardboard, and paper. If all these different things had been truly excited, if they had somehow taken in energy from the emanation, they might have been expected to give it back in characteristically different ways, but in fact they all gave rays with exactly the same penetrating power. A single substance, laid down by the emanation, would do precisely that.

If there was a deposit on the plates, then it should be possible to get it off. Rutherford found that a blast of air would not dislodge it (so it was probably not a dust). He could not flame it off (so it was probably not a "dew" of condensed emanation). He could not wash it off a platinum wire with hot water or with cold. He could not dissolve it in a strong alkali, nor with concentrated nitric acid. With dilute sulfuric acid, or dilute hydrochloric, it vanished in a few seconds. After that, he was quite interested to discover that the lost radioactivity reappeared on the bottom of the dish when he evaporated the acid, just as salt or sugar would be left behind from their solutions.

By September (this was 1899) Rutherford had finished with the emanation, and mailed off a paper to be published in England. By November he was ready with a paper on the excited radioactivity. The two were published in January and February, but long before that time, almost on the heels of Rutherford's ending, there had appeared a paper by the Curies on a radioactivity induced by the presence of radium. They knew nothing of Rutherford's work, of course, and the new radioactivity they found was so very short-lived, making such a great contrast to the permanence of radium and polonium, that they could not believe it came from any special substance. It seemed to them really excited, the result of some transfer of energy, perhaps by the rays of the radium, and the experiments they tried, looking at it in this way, were entirely different from those Rutherford had done. Perhaps the two radioactivities were alike, and perhaps they were not, though it was worth noticing that the Curies had seen nothing like an emanation.

The science of radioactivity was certainly not growing simpler.

5. Uranium X and Thorium X

As you may have noticed by now, science is a kind of free-for-all game that anybody can play at any time. When Röntgen discovered X-rays, he stimulated Becquerel to find the rays from uranium and J. J. Thomson to work out the theory of the ionization of gases. Becquerel's discoveries with uranium set the Curies to hunting out polonium and radium. Rutherford's interest in ions drew him on gradually to find the emanation from thorium and that curious thing it produced, which gave the excited radioactivity. What keeps the game going is publication, for although a man may publish to get the credit of a discovery, the profit to all the rest of the world lies in the information he makes available, information from which anyone may pick up an idea about something even newer to try.

It need not surprise you then that new players take it into their heads to join, and it was just as a new player that Sir William Crookes entered the study of radioactivity toward the end of 1899. He was a chemist who lived in London, where he had a private, consulting practice, and where he edited and published a chemical weekly—a picturesque person with a slender beard and white mustache waxed and twisted into long neat points. He was also a man of means who liked research and could afford to do it in a private laboratory fitted up at the back of his house.

As a chemist, he wanted to try his hand at extracting radium from pitchblende, and he knew how tedious the work of concentration was going to be. He knew that he must watch at every step, and gauge the radioactivity of every precipitate to be sure that the radium went always where he intended. He had been a pioneer photographer, and rather than tinker with such a cranky instrument as the electrometer, he decided to measure his radioactivities by a method he understood, with photographic plates. He knew what difficulties they would get him into. The blackening of a plate would depend upon other things than the exposure it got: the sensitivity of the emulsion which coated it for one, the strength of the developing solution for another, and none of them would be entirely under his control. All the same, they could be managed if for each plate of the series of his tests he put alongside a standard exposure from some material whose radioactivity he could guarantee.

Uranium seemed the proper choice for a standard material, but he thought it safest to use pure uranium, and he was chemist enough to know that the purity he wanted could not be bought. He laid in some pounds of uranyl nitrate of quite ordinary quality and set about purifying it himself. He put a couple of handfuls of it into a separatory funnel, (a bulb of glass with a stoppered opening at the top and a stopcock at the bottom), poured in some ether, and shook the mixture together. The uranyl nitrate began to dissolve in the ether, and as it dissolved its water of crystallization was set free. Since water and ether do not mix, the water scattered into tiny drops, which, as the shaking went on, washed out of the ether some of the uranyl nitrate and the greater part of the impurities. When things had gone far enough, Crookes stopped the shaking, let the heavier water settle to the bottom, drained it off through the stopcock and threw it away. Then he evaporated the ether to recover his purified uranyl nitrate, and he was ready for the next step.

This was a fractional crystallization. He dissolved his improved uranyl nitrate in the smallest possible amount of hot water, and let the solution cool slowly. As it cooled, new crystals of even purer uranyl nitrate began to form, leaving the impurities once more in the water, with which they could be drained off when the cooling was finished. After two or three repetitions of this process, Crookes was satisfied. His uranyl nitrate could now be considered pure. He was ready at last for serious work, but when he tested the pure uranium standard, his plates came out totally blank. Either the chemical treatment had spoiled his uranium, or he had managed to purify away its radioactivity.

A little strenuous testing convinced him that uranium was not easily spoiled. He tried the ether-washing on a fresh batch of uranyl nitrate, and found the radioactivity where he now expected it, in the waste water. Of course, there was uranyl nitrate in the waste water too, so the next problem was to find a chemical treatment to bring out the radioactive substance free from any mixture of uranium.

It was a simple treatment when Crookes found it. He dissolved some uranyl nitrate in water, and stirred in large doses of ammonium carbonate solution. At first the uranium precipitated, but as he went on, the precipitate dissolved, leaving only the least residue for him to filter off. It was brown and fluffy as it lay on the filter paper, and he recognized it as aluminum hydroxide colored with a little iron. There was no uranium in it, but on a photographic plate it produced beautiful darkening, in as little as five minutes.

There could be no doubt now. In that minor residue, alongside the commonplace aluminum and iron, lay the "impurity" to which the radioactivity of uranium really belonged. To judge from its chemical behavior it could not be polonium, and it did not seem to be radium. It was quite possibly new and provokingly odd, and Crookes expressed all his feelings of wonder and perplexity by naming it uranium X.

It was May 1900 when he finished and reported all these things to the Royal Society in London. In July his discovery was confirmed in a paper which Becquerel read before the *Académie* in Paris. He had borrowed a method of purification from André Debierne, a former student, who was now doing chemical work for the Curies, and it was hardly as effective as those which Crookes had used. Becquerel never achieved totally inactive uranium, but by persistence he did manage to push five-sixths of the radioactivity out of one hard-used specimen.

That was the summer Rutherford spent in New

Zealand, returning at last to be married after nearly five years away, and completing a trip around the world as he did so. That was the summer also when young Frederick Soddy turned up in Montreal looking for work. He was not quite twenty-three, from the southeast coast of England, a chemist who had taken his degree at Oxford something over a year before. The job for which he had come to Canada fell through, but during his stay in Montreal he was so mightily impressed by the magnificent laboratories that Sir William Macdonald, the tobacco millionaire, had built for McGill, that he was happy to be taken on there as Demonstrator (or laboratory instructor) in chemistry. As casually as this, he walked into a career, for Rutherford presently found him and enlisted him for the work on thorium.

Rutherford had radium to work with, too, for a German chemical firm had put it on the market in rather weak preparations, and radium was interesting since it was known by now that it had an emanation like thorium. This fact had been discovered by Professor Ernst Dorn at the German University of Halle, and it stimulated Rutherford's curiosity when he came back in the fall to see what the new emanation was like. He soon found that he could increase the flow of emanation by heating his materials, and if he heated them strongly, he could get an enormous burst of it, after which, especially in the case of thorium oxide, it was difficult to extract much more. This seemed to say that each solid preparation had about so much emanation locked up inside, and when that was driven out, it was gone for good. As Rutherford also discovered, both his radium and his thorium oxide when they were gently heated would give out emanation for hours on end, much more all told than the strong heating brought. This looked as though new emanation were being freshly produced, and it suggested something like a chemical reaction.

As 1900 turned into 1901, Rutherford and Soddy combined forces. (Rutherford could think up questions faster than he could find help in answering them. About the emanation, in particular, there were a number of things a chemist might find out.) Soddy was easily interested, and they drew up a set of five questions for him to investigate:

Did the emanation really come from the thorium, or was there some other, hidden substance which released it?

Was the thorium oxide permanently damaged when it was heated so strongly that it lost its power to give the emanation? Or was there a chemical treatment that would restore it?

What sort of a gas was the emanation?

Could very careful weighing on a sensitive balance show that a preparation which gave out emanation was losing weight? Or that a body gained weight when it picked up the excited radioactivity?

What chemical peculiarity of thorium made the production of emanation possible?

Some of these were easy questions and some extremely difficult. The fourth, for example, could be settled by simple arithmetic. An ordinary electrometer could detect 3×10^{-13} coulomb of electricity. (This is not a very large amount. Three thousand billion such charges would make up a whole coulomb, and this is no more electricity than will filter through a 60-watt light bulb in two seconds or flash through an electric toaster in a tenth of a second.) In the electrolysis of water, 10^5 coulombs must pass through a cell to set free a single gram of hydrogen (and so electrolysis can be an expensive process unless electric energy is more than ordinarily cheap). Hence the electrometer could detect an amount of electricity which it took only 3×10^{-18} gram of hydrogen to carry. It was quite clear then that the number of ions collected in an ordinary measurement corresponded to an unbelievably minute quantity of matter, and that the radioactive materials whose rays produced them must exist in an extraordinary degree of fineness, millions of millions of times below the range of the balance.

For the other questions there are hints of difficulties and of strange happenings in Rutherford and Soddy's final report, but what Soddy accomplished amounted to this. He tried a number of approved methods for separating thorium from a mixture of other elements and found that his purified thorium always gave out the emanation. That seemed to settle the first question. He tried a whole series of carefully planned chemical reactions to capture the emanation from a gas stream and lock it up in a solid compound. Every one failed, although among them all he should have been able to trap each one of the known gases. That made the emanation quite possibly a member of the argon family of inert gases, which Professor William Ramsay of University College, London, had been discovering over the last seven years. This disposed of the third question. For the second, Soddy found a strenuous but practical routine with which he could take the overroasted thorium oxide into solution in water, and then recover it as a freely emanating compound.

Here the oddities bobbed up again. It was not quite possible to bring the damaged thorium oxide exactly back to its original condition, but when Soddy converted it into thorium hydroxide, he did even better. When it was freshly precipitated from solution, thorium hydroxide gave quite as much emanation as the best commercial oxides, and, as it aged, it improved, until after nine days its output of emanation had become two and a half times as great. That was interesting, and so was another experiment in which Soddy divided the thorium in a solution of thorium nitrate between two precipitates, one of thorium hydroxide and the other of thorium carbonate. Weight for weight, the hydroxide gave twelve times as much emanation as an ordinary oxide; the carbonate gave almost none.

This was not only interesting but confusing, since it came about through an unconscious mistake, but the flurry of work it set in motion led Soddy soon to another curious discovery. The way he obtained thorium hydroxide was to dissolve thorium nitrate in water and add ammonia to bring down the hydroxide as a precipitate. This time, when he had filtered off the hydroxide, he saved the filtrate, evaporated it, and found a very scanty residue which contained no thorium and gave off quantities of emanation. Unfortunately, so did the hydroxide he had just filtered off.

At this it seemed best to abandon the emanation, and go back to straight measurements of radioactivity, where Rutherford felt more at home. Then at last they got an uncomplicated answer. Soddy went through the same operations again, and now there was a strong radioactivity in the thorium-free residue remaining in solution, while the radioactivity of the hydroxide precipitate turned out to be very much reduced.

In the midst of their work, one of them had run across Crookes's paper in the *Proceedings of the Royal Society,* and it was clear that what they had found was a thorium X. It was an "impurity" but, after their calculation of quantities, probably not an ordinary one. Their thorium nitrate was certainly not pure. Crookes had mentioned a highly refined thorium nitrate which could be bought in Germany, and it seemed easier to buy some of this material for their next experiments than try to clean up their own.

In any event, Christmas was coming, and this seemed a good place to break off the work. They had a paper nearly ready describing Soddy's systematic investigations. To this they tacked their latest discovery as a surprise ending and mailed it off to London, along with a letter to Crookes inquiring about the German thorium nitrate. Then they closed the laboratory for the holidays.

(A week or so earlier, the first Nobel Prize in Physics had been awarded to Röntgen for his discovery of X-rays.)

THE ATOMS TODAY

6. Thorium X and Transmutation

Over in Paris, during those same last months of 1901, Henri Becquerel had grown disturbed. There was no doubt that the radioactivity of uranium could be taken away by straightforward chemical processes, but neither was there any doubt that uranium was always radioactive. Everyone who had worked with it had found it so, no matter where the uranium came from, how it had been extracted from its ore, or what particular compounds had been studied. In all these various circumstances it was odd that no one had stumbled on even one pure, inactive specimen.

There was a logical way out of the contradiction. It was easy to purify away the radioactivity of uranium, but in the end uranium was always radioactive. Then the purified uranium must possess a power of reactivating itself. This was logical if implausible, and it could be tested experimentally, for Becquerel had carefully saved all his old preparations from the summer of 1900, each de-activated uranium and all the radioactive impurities. He had only to get them out and test them.

He did and found that every specimen of uranium, whether he had taken much or little away from it, had now regained the same, high level of radioactivity. By the same token, every radioactive impurity had gone dead.

This was both satisfying and confusing. Becquerel had always thought of radioactivity as an outpouring of stored-up energy, and its return would simply mean that the uranium had filled up again from the same mysterious source. On the other hand, a chemical operation had taken radioactivity away, and this told him that it had gone off with a particular variety of molecule. (Becquerel was too much of a physicist to speak of atoms.) Then its return must involve some kind of molecular change. He himself had shown that the rays from uranium were swiftly moving electrons. If electrons were tiny bits of matter (as J. J. Thomson was maintaining), then they might well be mixed up in the change he imagined.

So much speculation was too much for the Curies, and they published a strong criticism of Becquerel's ideas. To be general, to stick to energy in discussion, and not try to imagine any detailed processes—this seemed the only safe course until a good deal more knowledge had come to light. Moreover, radioactivity was a matter of atoms (that is to say of elements) without regard to the particular molecular combinations they entered. If Becquerel implied that there were atomic changes going on, they had seen no evidence of any. Over a period of months their radium preparations had shown no change either of weight or of spectrum. (This was an argument which Rutherford and Soddy had already refuted, but their manuscript was hardly yet in the hands of the London printer.)

Long before this, however, Becquerel had written to Crookes, asking him to check on his own purified uranium, and Crookes had passed on the word to Montreal when he had forwarded Rutherford's order for the purified thorium nitrate to Knöfler, the German manufacturer. So when Rutherford and Soddy came back to their laboratory in the new year of 1902, they found waiting for them Crookes's letter, Becquerel's paper, and the Knöfler thorium nitrate. They found also that their thorium hydroxide precipitates were thoroughly radioactive, and all their thorium X had gone dead.

It is plain that they had expected nothing of the kind before Christmas, although on looking back they may have felt that the odd changes in emanating power could have given them warning. In any case, they knew now precisely what to do. The Knöfler nitrate contained thorium X when they tested it, and this told them that it was no ordinary impurity. Using the Knöfler nitrate, they must separate the thorium and thorium X once more, and then follow out separately the changes in radioactivity which each of those substances showed. What they found brought them directly back to things Rutherford had seen two years before. The thorium X lost its radioactivity by a geometrical progression in time, dropping half of it every four days. The thorium hydroxide, starting with a small radioactivity, worked up to a steady value about four times as great, picking up half of what it lacked in the same four-day period. (Fig. 10)

If the curves were the same, then the explanation was the same. Within the thorium hydroxide, thorium X was being steadily produced, and once produced, it was losing radioactivity by the stand-

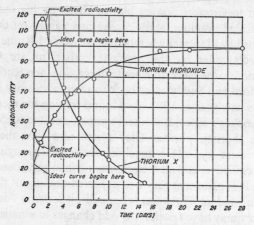

Fig. 10. THORIUM HYDROXIDE EXPERIMENT. *The declining curve represents the decay of radioactivity in the thorium X that Rutherford and Soddy extracted from thorium hydroxide. The rising curve represents the growth of radioactivity in the thorium hydroxide from the transmutation of its thorium into thorium X. As in the previous diagrams, the circles show actual measurements; the curves are ideal.*

ard geometric progression and at its own characteristic rate. At the beginning, when there was little of it, the half of the thorium X that ceased to be radioactive was a far smaller quantity than the amount the thorium hydroxide produced; the radioactivity of the thorium would steady off only when the loss of radioactivity had grown equal to the production of fresh material.

The problem was to imagine where the thorium X came from. It might come from nowhere, or it might come from the thorium. Both guesses were absurd, but the first was also incredible. If thorium X was a substance, if it was made of matter, then the only matter it could possibly come from was the thorium. Since the two were chemically different, they must be made of different kinds of atoms. To put it bluntly, thorium was an element, thorium X another, and the atoms of thorium must be steadily transmuting themselves into atoms of thorium X.

No other conclusion was possible. Rutherford and Soddy had first to convince themselves (against the Curies' skepticism boldly in print), and then all the rest of the world.

They were saying that the regrowth of radioactivity in the thorium hydroxide meant a regrowth of thorium X there. Then that thorium X should be as easy to separate as the original amount had been. They took an old hydroxide from which the thorium X had once been extracted, dissolved it in nitric acid, and poured in ammonia to precipitate the thorium again. In the filtrate they found the usual amount of thorium X. Twenty-four hours later they dissolved the hydroxide again and once more brought down the thorium. This time the filtrate contained a sixth of the usual amount of thorium X, just about what should have grown in the twenty-four hours of waiting. Six hours after that they tried again, and this time the accumulated thorium X was down to a thirtieth, again according to expectations.

The thorium X did grow, but this did not necessarily make it an atomic product. It might possibly be produced in an ordinary chemical change, and if so, then the rate at which it came into being should be affected by outside conditions. Chemical reactions occurring quickly in solution might be impossible to start in solid mixtures; warming would generally speed them up and chilling slow them down. This was not the way the thorium X behaved, as Rutherford and Soddy found in a whole series of tests. It grew, as its growing radioactivity showed, at precisely the same rate whether the thorium hydroxide was wet or dry, hot or cold, in the solid state or in solution.

Then if thorium transmuted into thorium X, thorium X might very likely be transmuting into the emanation. At least, it was true that thorium X released emanation in direct proportion with its own radioactivity.

Transmutation gave the simplest explanation for everything they had seen, but it was certainly a radical notion, and they used the greatest care in drawing up the paper in which they meant to propose it. In particular, they took pains with an argument based on energy. The new theory, they said, fitted very neatly with the ordinary notions about energy. When radioactivity had seemed perpetual, it had been necessary to imagine some way of pouring back into radium or uranium or thorium the energy that was always escaping in rays. Now, instead, you could suppose that an atom of thorium contained a certain store of locked-in energy. It transmuted to an atom of thorium X, and a part of that energy became available for release. It would trickle out slowly, and the rays carrying it away would gradually diminish as the available energy in the atom grew less and less. Then a new transmutation would unlock a new portion for the emanation to spend, and the transmutation of the emanation, quite likely, still another portion to supply the excited radioactivity. In this way, you could account for "permanent" radioactivity as a very slow process of spending. The number of radium or thorium atoms transmuting would be so small, that they would make no perceptible change in the number that were left.

There were still a few perplexities to be worked out, and the worst of them was the persistence of the radioactivity of thorium. Crookes had swept his uranium free of uranium X, and with it had gone all the radioactivity. Soddy was able to extract thorium X, and apparently all the thorium X, from thorium, but the hydroxide precipitates always kept about a third to a quarter of their original activity. Yet this, they thought, with some other details, would all clear up in time.

Then, after they had discussed energy and edged their way around the unsolved puzzles, they were ready to spring the trap of their logic. First (they said), radioactivity was an affair of atoms—and this was the Curies' firmly stated opinion. Second (they went on), it was "the manifestation of a *special kind of matter* in minute amount." (This idea was Rutherford's particular contribution.) Therefore (and the trap snapped shut), radioactivity must be the "manifestation of sub-atomic chemical change."

With this they had done their best. They closed the paper with a few more paragraphs of persuasion, and mailed it to the Chemical Society of London. For extra assurance, Rutherford wrote to Crookes, since he had been so friendly in the matter of the thorium nitrate, and asked Crookes's help to move the paper across the editor's desk.

7. Rays and Transformations

The new theory was plausible and promising, but to be strictly honest, Rutherford and Soddy had given it only one thorough test, on the transformation of thorium into thorium X. The production of uranium X would probably follow the same rules, and Soddy set about to repeat Crookes's experiments, only to fail completely. The uranium remained unaltered, and there was no sign of any uranium X. (It was strange how regularly each new experiment dealing with uranium went wrong.)

This was impossible. Crookes was a first-rate chemist, and Soddy at least reasonably competent, but it was not until his desperation had driven him to an exact copy of Crookes's procedure that he saw the cause of the trouble. Crookes had used a photographic plate to detect radioactivity, and on a photographic plate Soddy, too, could show that uranium gave no rays while uranium X had them all. But measuring ionization, as he had learned from Rutherford, Soddy found rays only with the uranium, and none with the uranium X.

Back at Cambridge, when he had been studying the ionization produced by the Becquerel rays, Rutherford had noticed that the rays from uranium were a mixture of two different kinds, and had labeled them with the completely meaningless names of alpha and beta rays. The alpha rays produced enormous ionization, but they had so little power of penetration that a single sheet of paper stopped them. The beta rays penetrated about as X-rays did and, in contrast, ionized rather weakly. The paper wrapping which had screened the ordinary light from Crookes's plates had stopped the alpha rays but let the beta rays through, while Soddy's ion-collector had responded to the alpha rays and given no indication of the betas. It looked then as if uranium gave alpha rays, and uranium X gave beta rays only.

Now Thomson had shown that the cathode rays were streams of high-speed electrons, and Becquerel had discovered a little later that the penetrating rays of radium and uranium were electrons of just the same kind. The cathode-ray electrons generated X-rays when they struck against the glass wall of their tube, and it was generally assumed that when beta-ray electrons were stopped by collisions inside the lumps of radioactive powder from which they started, they generated X-rays too. Very plausibly these X-rays made up the soft, ionizing radiation to which Rutherford had given the name of alpha.

At that very moment at McGill, a young physics instructor named A. G. Grier was finishing an investigation of the connections between the alpha and the beta rays of the different radioactive substances. (Grier was properly an electrical engineer, but Rutherford had annexed him as easily as he had taken on Soddy.) He had developed a special ionization chamber for beta rays alone, and with it he confirmed what Soddy had already guessed, that uranium gave nothing but alpha rays and uranium X nothing but beta. Then with Soddy doing the chemistry, he tackled the thorium products, and found that although thorium X gave rays of both kinds, thorium was like uranium and gave alpha rays only. Most of his winter's work was knocked apart now, for if either ray could show up without the other, there was obviously no connection between them. But the new discovery was fascinating enough to make up.

Soddy was a chemist. For the sake of the reputation he had still to make, the two long papers on the emanation and the production of thorium X had gone into a chemical journal. That meant that no physicist would ever see them, and so, as the spring of 1902 wore on into summer, the papers were redrafted for the leading physical journal of England, which bore an old-fashioned name, *The Philosophical Magazine.* They began with the second paper, which was far more important, and it was perhaps six weeks later that they set about putting the interesting parts of the first into shape.

It was a slack season, when classes were over, when most of the old experiments had been wound up and no new ones started. They had been busy, and now they were relaxing. It is at times like this that ideas begin to shift, that old notions take on a new appearance, and snippets of information that had seemed quite independent turn out to be closely related. Something of the kind happened to Rutherford and Soddy. When they had finished describing their early emanation experiments, they changed the subject abruptly, and closed the paper with a new version of their transformation theory. It did not seem like an enormous change, but it was amazing how much more power it gave the theory.

Previously they had pictured the transmutation process like this: There was an alteration within an atom, some change in the way it was put together, which made a change also in its chemical behavior. This change brought to the surface, as it were, a certain amount of energy, which the transformed atom proceeded to spend by pouring out a long stream of slowly dying rays. Now, they suggested that the energy came in a single burst of rays, within the very instant in which the atom transformed itself. The giving-out of rays became an intimate part of the act of transmutation.

When they had thought that the rays signaled the presence of atoms already transformed, they had been puzzled to explain how unaltered uranium and thorium could give out alpha rays. Now these rays became necessary as the outward sign of their acts of transformation. Each alpha ray from thorium told of an atom which had changed to thorium X, each ray from thorium X, the appearance of an atom of emanation. The rays from the emanation marked its transformation into the deposited matter of the excited radioactivity, and the rays from the excited radioactivity marked still further changes within that solid deposit.

It followed now that the intensity of the rays gave the number of transformations that were actually going on. Then it was right to expect, as their experiments had shown them already, that where the radioactivity of thorium X was high there would be a rapid production of emanation, and where the emanation was active, large quantities of the excited radioactivity would appear.

Finally, the new theory made perfect sense of the geometric progression in the decay of radioactivity. There was no way to change the rate of this decay by any change in outside circumstances, and this showed even more plainly in Pierre Curie's experiments than in those Rutherford and Soddy had done. This meant that there could be no cooperation between one atom and another in bringing out the rays. If there had been, warming the atoms to bring them more often into contact should have changed their radioactivity, and so should locking them up in crystals to hold them apart. Since the giving-out of rays was an act which each atom managed separately, and since this act was also the act of transmutation, then a little calculus was enough to show that the transmutations and the ray-giving must occur by the geometric progression.

A theory is useful when it accounts for things already known, but it is useful also when it suggests new things to look for in the future. The new transformation theory managed to do both, for it suggested an untried experiment, which Rutherford found irresistible, to discover what the alpha rays really were. As long as they seemed a side-effect to the beta rays, they had hardly been worth investigating, but he knew now that they were not a side-effect, and that they played a part in the very first transformations of both uranium and thorium. What was more, he saw that they need not be X-rays. In fact, to carry off the energy of transforming atoms in a single burst, they might better be corpuscular—that is to say, some sort of flying, sub-atomic particles like the beta-ray electrons.

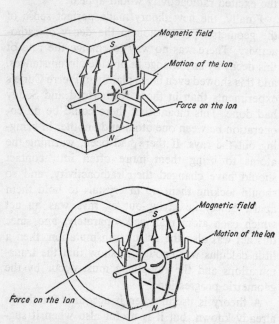

Fig. 11. Ion Behavior in a Magnetic Field. *An ion moving through a magnetic field experiences a force perpendicular to the field lines of force and to the ion's direction of motion. The direction of the force depends on whether the ion's electrical charge is positive or negative.*

Since the excited radioactivity would collect on negatively charged plates, the atoms giving it out must be positively charged themselves. These atoms came from atoms of emanation, each of which had lost an alpha ray, so it seemed likely that the alpha-ray particles had carried off the missing negative charges. If they did, then these swift-flying, negatively charged particles could be handled by a magnetic field.

When an electric charge is carried through a magnetic field, a crosswise force develops on it perpendicular to the lines of force of the magnetic field and perpendicular also to the direction in which the charge is moving. Thus a charged particle must swerve as it crosses a magnetic field, and the amount by which it swerves will depend (in a somewhat complicated way) upon the charge that it carries, the mass that it possesses, the speed with which it travels, and the strength of the field it crosses. (Fig. 11)

A little more than two years before, Pierre Curie had tried the effect of a magnetic field on the rays from radium. He had found that the beta rays bent sharply as the electrons composing them swerved, and the alpha rays sailed straight ahead. To be strict in interpretation, his experiments had shown only that the alpha rays did not swerve much. Rutherford made the investigation more stringent.

He put some radium at one end of a narrow corridor and set his ion-collector at the other. On either side of the corridor he placed one of the pole pieces of the laboratory electromagnet. Then he tested to see whether more rays got through with the magnet turned off than when it was on. The only trouble was that if he made the corridor very narrow, not enough rays came through for the ion-collector to work on, and if he made it wide enough to get a little ionization, there was plenty of room for an alpha particle to swerve without hitting its walls.

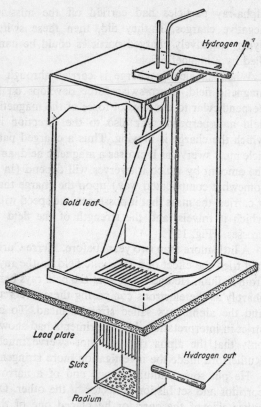

Fig. 12. RUTHERFORD'S ALPHA-RAY DETECTOR.
Hydrogen blown in the tube at the top swept back
"emanation" from the radium, and the lead plate
shielded the electroscope from the radium's pen-
etrating rays. The angled terminal projecting from
the electroscope could be pivoted away from the
metal plate to prevent electrical leakage after the
plate and gold leaf had been charged. Ionization
produced by alpha rays passing through the box of
slots at the bottom of the apparatus discharged the

One obvious improvement was to increase the strength of his rays. With Pierre Curie's help, he persuaded a French manufacturer to sell him a mixture of barium and radium chlorides richer than they usually put on the market. Another was to provide paths which would be both wide and narrow, and this Rutherford managed by inventing a little metal box, broken up into slots by a long set of parallel plates, which made a great many, very narrow corridors side by side. A third was to increase the sensitivity of his ray detector, and this he did by replacing his usual ion-collector and electrometer with a gold-leaf electroscope of a new pattern, directly above the box of slots.

(The electroscope was no more than a narrow metal plate supported by an insulating plug and with a strip of gold leaf hinged on near the top. Any charge put on the metal plate was shared by the leaf as well, and since like charges repel each other, the very light gold leaf would be forced out at an angle to the plate. Ions formed by the alpha rays coming up through the slots would be attracted to the leaf, and as they reached and discharged it, the leaf would gradually sink. Rutherford could watch its fall through a long-range microscope, looking in through a glass window in the case, and could time the edge of the leaf as it moved past a numbered scale in the eyepiece. Thus he could work out its speed, which was, of course, a measure of the intensity of the rays. Fig. 12.)

---

gold leaf. To test the effect of a magnetic field on the alpha rays, Rutherford arranged the apparatus with the box of slots between the poles of an electromagnet.

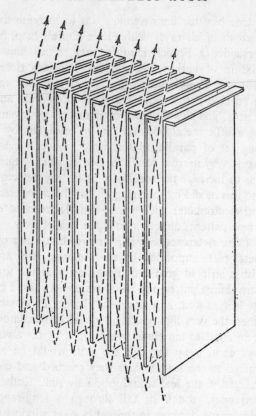

Fig. 13. A RUTHERFORD TRIUMPH. *With this device, a development of the box at the bottom of the apparatus shown in Fig. 11, Rutherford discovered that alpha particles carry a positive charge. A magnetic field made the alpha particles swerve as they traveled from the radium through the corridors of the slotted box toward the electroscope's gold leaf. To find in what direction they swerved—and therefore whether their charge was positive or negative—he capped the upper ends of the cor-*

When the little box was in its place between the pole pieces of the magnet, and the new radium was spread across its bottom, the gold leaf sank in a very satisfactory style as long as the magnet was turned off. When Rutherford turned it on, it was plain that the leaf dropped more slowly, and he knew that the alpha rays were bending a little. That was all he could manage with the strongest current he dared use in the Physics Department's magnet, and at this point Owens came to his rescue. They partly dismantled the largest dynamo in the Electrical Engineering Laboratory, replacing its pole pieces with smaller ones, which were closer together. Then they put in the box, turned on the current, and the gold leaf stood stiff. The alpha rays were all being thrown into the corridor walls by the magnetic field, and they could be nothing except swiftly moving, charged particles. On days like this, when everything went well, Rutherford could be heard all up and down the hall in a triumphant chant of "Onward, Christian Soldiers."

The next thing was to make sure of the sign of their charge, whether it was positive or negative. In a particular magnetic field, positive and negative charges would be forced in opposite directions, so all that Rutherford needed was to find which way the alpha rays curved. To do this, he had a box made up in which each slot carried a little cap, closing one side and leaving the other open. Now a weaker magnetic field could stop the ionization where the gold leaf hung, but only if it threw the

----

ridors, reducing the size of the opening. If the particles turned left, they would hit the caps. If they turned right, they would pass through the reduced openings.

alpha particles in under the caps. If they swerved
the other way, toward the openings, they could
still come through and the gold leaf would continue
to drop. Rutherford tried out the new box and, to
his intense surprise, found that the alpha particles
were positively charged. (Fig. 13)

(This meant that as the atom of emanation
transmuted, it lost positive charge with its alpha
particle, but since a number of negative electrons
might be torn loose in the process, carrying away
a greater total charge, the atom of the new element
might still come out positive.)

The alpha particles were astonishingly hard to
move, so either they were traveling at enormous
speeds or else as (compared with electrons) they
were enormously heavy. It was not possible to dis-
entangle these two influences with the magnetic
field alone, but if Rutherford could also manage to
bend the path of the flying particles with a strong
electric field between two charged plates, then he
would have the mathematical means for doing it.
The new experiment was easy enough to design,
for it was only necessary to make the box of hard
rubber to insulate the dividing plates, and then ar-
range connections so that the plates could be
charged plus and minus in alternation. The problem was to make it work, for the alpha particles
ionized very heavily, and the ion currents to the
charged plates had a way of breaking over into
sparks before the electric field was strong enough
to do much good.

In the end he had to be content with a guess.
From the length and width of the slots, and the
rate at which the gold leaf slowed down, he could
tell something about the curve which the alpha par-

ticles followed. With the magnetic field he had more definite knowledge. Running out the arithmetic, he found the speed of the alpha particles to be 2.5×10^9 centimeters per second, or about a tenth of the speed of light. The ratio of the charge each particle carried to its mass was around 6000 in the units he chose to use (abcoulombs per gram).

This told more than you might think, for just this same ratio of charge carried to the mass of the carrying particle appeared in the electroplating experiments of electrolysis. When 9650 abcoulombs of electricity passed through the proper solution, it would plate out 108 grams of silver, 32 grams of copper, or 1 gram of hydrogen. Since all the ions in a solution ought to be of the same kind, if you divided this total electrical charge carried by the mass of material plated out, you should have the charge-to-mass ratio for the separate ions. This comes out, if you do the arithmetic, to a little less than 90 for silver, almost exactly 300 for copper, and 9650, or under 10,000 for hydrogen. As these figures (and the arithmetic) make plain, it is a number that grows larger the lighter the ion is. For electrons (in the fall of 1902) the charge-to-mass ratio was known to be something like 10^7, and this very large value was one of the bits of evidence that the electrons must be sub-atomic particles—that is to say, only small fragments of atoms.

It was easy to see that the alpha particles were nothing like electrons, but that they must be at least as heavy as ordinary light atoms. Since Rutherford's value of 6000 for their charge-to-mass ratio was only a guess, there was no point in pushing the argument, in trying to decide whether they were atoms of one of the known elements or some special body never before recognized. The bare facts, just as they stood, were interesting enough.

It was only these facts that Rutherford reported when he wrote up his experiments. Nevertheless, it was plain that this positively charged, atom-sized alpha particle was exactly what the new version of the transformation theory required. An atom shooting out one of them must certainly become very different.

8. Transformations and Energy

Was the transformation theory true? This seems like a reasonable question to ask here, but it would be better to ask (as we have already suggested) whether it was useful. We tend to feel a strong attachment to things we call true, to find it difficult to give them up or even to argue about them calmly when they are challenged. In science, as we have learned by hard experience, theories do not always last, and when a theory no longer helps, it had better be abandoned for one that does.

The transformation theory had done very well so far, and through the winter of 1902–3 it continued to prove its value as Rutherford and Soddy pushed on their investigations.

Now that they knew which rays were which, Soddy separated some uranium X from uranium, and for four months they followed the beta rays of the two specimens. Just like thorium X, the uranium X lost its radioactivity, and just as with thorium, the activity of the uranium seemed to pick up as the increasing beta rays signaled the presence in it of more and more uranium X. The only difference was the time-scale; the uranium X took twenty-two days for its transformation to be half-completed.

The emanations came a little nearer to seeming real (there is nothing more vague than an imperceptible gas) when Rutherford and Soddy showed that both the radium and the thorium emanation could be condensed in a sufficiently cold spot. Machines for making liquid air had just come on the market, Sir William Macdonald had been persuaded to buy one for McGill, and the instant it was working (days before it was formally dedicated), Rutherford and Soddy commandeered it for their own purpose. The experiment they tried was to blow air mixed with emanation through a pipe coiled up in a corkscrew twist and test the ionizing power of the mixture at the farther end. Since the ionization stopped whenever the corkscrew was dipped into liquid air, and started again when the corkscrew warmed up, it was reasonable to conclude that the emanation was condensing there.

Using the new radium sample, Soddy was able to get enough radium emanation to test its chemical behavior. Since it refused all chemical combinations, it, too, must belong to the argon family.

The production of emanation was the one radioactive transformation that outside conditions seemed to influence. Radium chloride gave out more emanation when it was warmed or in solution than it did in its ordinary state, and thorium oxide showed something of the same effects. Yet this did not necessarily mean that the production of emanation was altered. It might very well be produced at a perfectly steady rate, and the only thing that moisture or heating could change would be its release from the solid powder within which it was formed. This was a question they knew how to investigate now.

Solid radium chloride gave off no emanation. When it was crystallized from solution and left to stand, if the emanation was produced inside it at a steady rate, the amount of emanation it held imprisoned would grow toward a fixed limit by the familiar geometric process of gain. As Rutherford and Soddy had now found, the emanation needed four days to lose half its radioactivity, and so after about a month the radium chloride should be holding as much emanation as it possibly could.

It was not hard to find how much emanation this was. They took a stoppered flask, put water in it, dropped in the radium chloride, and blew air through the solution and into a gas-holder until they were sure that all the imprisoned emanation had been swept away. Then they stopped up the flask tightly, set it away and measured the emanation they had extracted from it by the ionization its rays produced. An hour and a quarter later they blew air through the flask again, and once more assayed the emanation it carried off. It came to the exact amount which the transformation theory predicted on the supposition that radium turned into emanation at a steady rate. The production of radium emanation (at least) was just like any other radioactive transmutation.

So the winter passed, and with the spring there came an offer to Soddy of a position in Ramsay's laboratory in London. (Professor Ramsay had now become Sir William, by the way.) Before the partnership dissolved, Rutherford and Soddy wrote one last paper, summing up all they had found so far about radioactivity, and ended it with a little atomic arithmetic.

If you know the mass and the speed of a flying body, you can calculate the energy which its motion gives it. Rutherford had just estimated the speed of the alpha particles from radium, and the mass each had could be calculated very plausibly from their charge-to-mass ratio, using Thomson's value of 2×10^{-20} abcoulomb for the charge which a single ion carried. This meant that each separate alpha particle took off with it something like 10^{-5} erg of energy. (An erg is the amount of energy which is spent in hoisting a mosquito up half an inch.) From the kinetic theory of gases came a calculation giving 10^{20} as the number of atoms in a gram of radium, and Rutherford and Soddy knew now of five successive transformations which each radium atom would go through. Putting these all together, they got a figure of 10^8 (or a hundred million) gram-calories for the heat a single gram of radium would give out before it had transmuted itself into quiet. (A gram-calorie is the amount of heat needed to warm a gram of water through one Centigrade degree, which works out at 42 million ergs of energy. The calorie used in nutrition is a thousand times as big.)

Granted that the whole calculation was made up of guesses, 10^8 gram-calories was out of all comparison greater than the 4000 produced by the fiercely burning flame of hydrogen in pure oxygen when a gram of water was formed. Radioactive changes involved quite a good deal more energy than the molecular combinations of ordinary chemistry.

This was not the end of the arithmetic, however. There was another line of calculation, a little more involved, which started with the total ion current which the emanation from their new radium sample could produce. From this they figured that a gram of radium which imprisoned its emanation, and so kept the matter of the excited radioactivity within its bulk, must be giving out energy at the rate of 2×10^4 ergs per second or 15,000 gram-calories per year.

It was quite plain that if radium was spending energy at this rate, even the hundred million gram-calories they had allotted it would not last very many thousands of years. The radium which was extracted from pitchblende must be far younger than the ore in which it lay.

If radium transmuted this quickly on the geological scale of time, then it was clear why so little of it had accumulated in the ores from which it was extracted. Nevertheless its transmutations must exhaust themselves in the end, its atoms must come at last to the stable existence of ordinary elements, and the stable end-products, whatever they might be, must lie in the ore beside it. Here Rutherford and Soddy pointed out the curious fact that helium, the lightest of Ramsay's family of inert gases, was always found imprisoned in radioactive minerals.

Then they shifted to a wider speculation. Radioactive atoms were releasing energy, and releasing it in enormous amounts. There was no reason to think, however, that these atoms of radium or thorium or uranium were very different from the atoms of ordinary elements, or that all their internal energy was spent in their rays. It was quite likely that stable atoms contained also their store of locked-in energy, and if sub-atomic processes existed to unlock it, then one might account for even the prodigious outpouring of energy by the sun.

It was a long way to go on a little arithmetic, but at least they had not exaggerated, and this was made plain before their figures could appear in print.

In Paris, radium was piling up. On July 21, 1902, when Rutherford and Soddy's transformation theory was hardly off the press, Marie Curie had announced that its atomic weight was 225, and this fitted it very neatly into the Periodic Table in the space below barium, and on the same row with thorium and uranium, just as its chemical behavior required. Alongside this specimen, which was absolutely pure, the Curie laboratory held others in which radium and barium were still mixed.

With one of these, which weighed about a gram and had perhaps one part of radium to five of barium, Pierre Curie made an interesting discovery. The radium sample was warmer than the air around it. With the help of a young research assistant named Albert Laborde, he began a careful investigation.

Comparing the radium-barium chloride mixture against an equal weight of pure barium chloride in exactly the same surroundings, they found that the radium sample kept itself a degree and a half warmer than the other. Comparing the vial of radium against a small coil of electrically heated wire wound into the same size, they began to calculate how much heat it must evolve to do this. Then for extra assurance they checked the measurement in a Bunsen calorimeter, where the heat of the radium could melt a little ice, and the expansion as the ice turned to water told exactly how much of the ice had gone. The figure they came out with in the end was 14 gram-calories per hour.

This was little enough, but there was not too much radium in the sample they had. On a reasonable estimate of what there was, they were willing to say that a gram of pure radium would produce about 100 calories per hour. This comes out, if you try, at nearly 880,000 gram-calories a year, or almost 60 times what Rutherford and Soddy had just calculated. Perhaps they had been conservative in their arithmetic; if so, the facts of the matter only added force to the arguments they had drawn from it.

9. Radium and Helium

Helium was a thoroughly romantic gas. The first
hint that it existed had come in 1868, during an
eclipse of the sun, when a spectroscope, catching
the light from a prominence at the very edge of
the sun's disc, had flashed up a yellow line, very
close to the familiar line of sodium. Later, when
the French astronomer Pierre Jules César Janssen
had found a way to make the prominences stand
out, even against the white background of light
from the sun's body, he discovered that the sun's
yellow and the sodium line were not quite at the
same place. The yellow line, as it turned out, was
one of a group of lines of different colors which all
changed brightness together as the prominences
flared up or died away, and so seemed to be related. The curious thing was that none of the
known elements gave precisely such a group. The
astronomer Norman Lockyer and the chemist
Edward Frankland formed a partnership to search
them over, but after the most careful hunting they
had to admit that they could find no element on

earth to which these lines might belong. Yet an element to produce them seemed to exist in the sun, and for this reason they named it helium.

Then in 1895, when argon, the first of the inert gases, was new, it was pointed out to Ramsay that when a mineral named cleveite was dissolved in acid, it released a gas which might possibly be argon. He found the gas easily enough, with Crookes's help he got its spectrum, but it turned out to be not argon but helium, the sun-element, never before identified on earth.

Now Rutherford and Soddy were suggesting that helium might be one of the end-products of radioactivity, one of the elements that uranium and radium were transmuting themselves to. In fact, since uranium and radium and helium were all found together, they were very probably related, and when Rutherford and Soddy separated in the spring of 1903, they divided the problem of relationship between them. Soddy would try to grow radium from uranium, and Rutherford to grow helium from radium.

In the German city of Braunschweig (which becomes Brunswick when it is spelled in English), at the quinine factory of *Buchler und Compagnie,* there worked a chemist named Friedrich Giesel. He had taken the hint from Marie Curie's first paper, and had begun to explore the residues from a uranium refinery. He had discovered for himself both polonium and radium, though always a little behind the Curies, and only recently he had found that the bromides worked faster than the chlorides in the fractional precipitations which separated radium from barium. By 1903, he was offering for sale at the quinine works very pure radium bromide at very reasonable prices.

Ramsay had bought some, and as soon as Soddy arrived from Montreal, he set him to work on apparatus for getting the spectrum of the radium emanation. This was perfectly reasonable. Ramsay had shared the discovery of argon with the physicist Lord Rayleigh, but the other gases of its family —helium, neon, krypton, and xenon—he had run down by himself. He knew how to handle minute bubbles of gas and how to produce their spectra; Soddy had experience with radium. Between them they could manage this new, inert, radioactive gas, which Rutherford had named emanation.

Unfortunately, the experiment went badly; the only spectrum they saw belonged to carbon dioxide, which had got in by some accident of contamination. They froze that out with liquid air, and of course condensed the emanation also. Then, as Soddy was quite confidently expecting, the famous yellow line of helium shone out. Rutherford was on vacation that summer and, as it happened, in London on that particular day. He had bought some of Giesel's radium bromide for McGill, and he was more than willing to lend it to Ramsay and Soddy for a second try. This time they saw the full half-dozen helium lines, which run across the spectrum from red to violet.

There was only one possible interpretation. Helium was one of the gases which combined with nothing. In no way could the radium have brought it along from the original ore, in no way could the radium have captured it from the air. The helium which glowed under Ramsay's sparks must have come into being as a gas within the solid crystals of the radium bromide, and to be imprisoned there must have been formed after those crystals had solidified from their last purifying solution.

Radium was an element now, an undoubted element with its own chemistry, its own spectrum, and its own atomic weight. Helium was an element too, on exactly the same terms. Then here was clear and straightforward evidence that helium was formed from radium, that one element did transmute into another. Who could disbelieve after this the ideas that Rutherford and Soddy had proposed so cautiously only fifteen months before?

10. Heat and the Alphabet
of Radium

When Rutherford left London he had gone to Paris, meeting the Curies on the day on which Marie Curie's doctoral thesis had been accepted at the Sorbonne. Then he had settled in Wales, where it rained continually, and finally returned to Montreal with the thirty milligrams of radium bromide he had bought from Giesel. There the first thing he tried was to measure its heat-production, and find the reason for the discrepancy between his estimate and Pierre Curie's measurements. While this might mean only that he and Soddy had been cautious in their guesses, it might mean also that they had misunderstood what went on, and that the radioactive substances were spending energy in something else beside their alpha particles.

What Rutherford wanted to do then was to see how closely the heat which radium produced matched up with the ionizing rays it gave out. He formed a partnership with Howard Barnes, a specialist in the heat measurements which Hugh L. Callendar, Rutherford's predecessor had introduced at McGill. They took the radium bromide and heated it strongly to drive off all its imprisoned emanation, condensing the emanation with liquid air into a little glass vial, and sealing it hermetically in place. Next, in alternation, they measured the heat produced by the radium-without-emanation and the emanation-without-radium. For the first few hours, the heat production of the radium fell off as the active deposit which the emanation had left there decayed (this was the "excited radioactivity" of Rutherford's earlier days). During the same time, the heat from the emanation tube increased as a fresh active deposit was laid down on its walls. Then the newly forming emanation within the mass of the radium bromide began to catch up, the steady decay of the separated emanation began to show. In the emanation tube the heat production died down geometrically, in the radium bromide it rose in the same way, in both with the half-value period of four days.

This was precisely what the radioactivities of these two samples would do, as measured by the alpha particles they sent out. Beyond any question then, the alpha particles were the chief carriers of the radioactive energy, and this energy was released from the atoms at the instant of their transmutation. It was also most satisfying for Rutherford and Barnes to see that while the heat production in one tube was going up and that in the other was going down, the sum of the two remained steady. The new emanation imprisoned in the radium bromide exactly replaced what had transmuted away in the other tube, just as the transformation theory required.

(About the time these measurements were finished, in December 1903, it was announced in Stockholm that the Nobel Prize in Physics for that year would be divided between Becquerel and the Curies.)

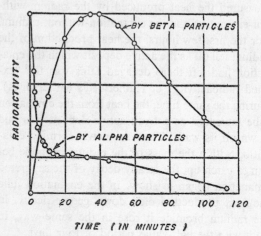

Fig. 14. ALPHA- AND BETA-PARTICLE RADIOACTIVITY *from the deposit laid down from radium "emanation."*

Next Rutherford and Barnes turned to the active deposit which the emanation of radium laid down. Here again, the production of heat followed exactly the alpha-particle radioactivity, and this, as Rutherford had known since he bought his first radium, died off in a rather complicated style. There was a geometrical decay at the beginning with a half-value period of three minutes. There was another at the end which needed about thirty minutes to drop by half. In between, for some twenty or thirty minutes, the radioactivity hung constant. (Fig. 14)

There were plainly two different substances involved here, the three-minute one at the beginning (which the emanation evidently transmuted into) and the thirty-minute one which showed up at the end. It could also be deduced, if you looked at it carefully, that there was a third substance between those two. The thirty-minute substance did not grow as rapidly as the three-minute one disappeared, there was that constant stretch which separated them, and this must represent a time when the first was slowly going and the last was still building up.

To follow the beta-particle activity made the argument a little clearer. At the start, an active deposit gave out no beta particles at all, evidently the three-minute material gave alpha particles only. If the element the emanation transmuted to was half-gone in three minutes, then the element it produced, the second element in the chain, must be half-formed in three minutes. It might be expected then that the beta-particle activity of the deposit would rise as rapidly as its alpha particles died away, but in fact it rose more slowly, and only gradually settled into the thirty-minute decay.

From all this it seemed necessary to conclude that the first element in the active deposit, the one the emanation turned into directly, gave alpha particles but no beta particles and was half-transformed in three minutes. The second element gave neither alpha nor beta particles but, transmuting quietly, was half-gone in thirty-four minutes. The third element in the series gave both alpha and beta particles and did half its transformation in twenty-eight minutes.

Pierre Curie had been very slow to accept the transformation theory. He was disturbed by the transitory kinds of matter it called for, and he much preferred to explain temporary radioactivity by a process of the transfer of energy. What changed his mind as much as anything was his own success in repeating Ramsay and Soddy's experiment, and seeing helium grow within some of the Parisian radium. (This had not been altogether easy, and he had needed help from Sir James Dewar at the Royal Institution in London and Henri Deslandres of the Astrophysical Observatory at Meudon to get a spectrum which was not masked by contaminating gases.)

Once he had accepted it, however, he began at once to measure up and analyze the alpha-particle activity of the deposit from radium emanation, in very much the same way as Rutherford. The emanation, as he put it, turned into a substance A with a half-value period of 2.6 minutes, that turned into a substance B with a half-value period of 21 minutes, and that into a substance C with a half-value period of 28 minutes.

The letters were very convenient for telling one element from another, and by the summer of 1904 Rutherford had taken them over, naming the three substances in order radium A, radium B, and radium C.

So far we have stuck to verbal arguments about the nature of these three elements, but the mathematical arguments were even more convincing. It was not difficult to set up equations ("differential equations" in the language of mathematics) to assess the growth of a radioactive element as the one before it transformed, its disappearance at the same time into the next element of the chain, and to balance one of these processes against the other. Once the equations were set up, they could be solved, and curves could be drawn from their solutions to show how much of an element might be present at any particular time. Rutherford might assume, to take a simple case, that he started out with nothing but radium A. From that beginning he could work out three curves to show how rapidly the radium A transformed away, how rapidly the radium B grew up from it, and how slowly the radium C followed after. (Fig. 15)

He had supposed that radium A and radium C were the only two which gave alpha particles. If their particles were equal in ionizing power, then if he added together the curves he had just drawn for radium A and radium C, the composite curve would give the whole alpha-particle decay for the active deposit. What came out was roughly like the curves he had plotted from his earlier experiments, and if you look carefully, you can see the reasons for the difference. The hump in the radium-A-plus-radium-C curve is much higher than in the experimental one, and this means that in those earlier measurements the radium C was ionizing less strongly than the radium A. On the experimental curve, the hump comes earlier in time, and this would mean that there had been a long period of collection of the active deposit, so that the radium C had begun to accumulate before the measurements started.

The active deposit of radium fell dead within a few hours, and at first it had been supposed that

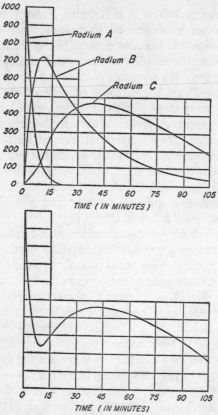

Fig. 15. TRANSFORMATION PREDICTIONS. *The curves of the upper graph show the calculated number of atoms of radium A, radium B, and radium C that would exist at any moment in a transformation process starting from 1000 atoms of radium A. Rutherford obtained the curve in the lower graph by adding together the radium A and radium C curves in the upper diagram. It should give the alpha-particle radioactivity of the active deposit.*

102

this was the end of its radioactivity, but then the Curies had noticed that old deposits became active again. With radium A, B, and C settled, in the early months of 1904, Rutherford began to investigate this effect. He had some fairly rich deposits in the emanation tubes of his heating experiments; he dissolved them out, and began to follow what they did. From the new material came both alpha and beta particles in a radioactivity which decayed very slowly if it did not actually hold steady.

It might have been polonium, for since polonium was extracted from the same ores as radium, it was possibly descended from it. Oddly enough, this rather "iffy" statement was the only certain thing which could be said about polonium.

Marie Curie had extracted it along with bismuth, and from her account polonium gave only alpha particles in quite a long-lived radioactivity. Giesel (who had prepared the pure radium bromide which Rutherford was now using, as well as that which had shown Ramsay and Soddy their first spectrum of helium) was accustomed to extract polonium with bismuth also, but his material gave beta particles and had rather a short life. Then there was the radiotellurium which Professor Willy Marckwald of the University of Berlin had found in the Joachimsthal residues when the chemical firm of Dr. Richard Sthamer had commissioned him to search them for something of commercial value. Like the Curies' polonium, it gave alpha particles only, and Marckwald claimed that it had a permanent radioactivity. He had gone to some lengths to prove that it was not bismuth (and so was not likely to be polonium), his most convincing demonstration being that when a stick of bismuth was dipped into a solution containing radio-tellurium, the radiotellurium promptly plated out on the surface of the bismuth. (An iron nail will do the same trick, coating itself with copper when it is dipped into a copper-bearing solution. It is a sign that the dipping metal is a good deal more apt to dissolve in water or weak acids than the one that plates out.)

Marckwald's was the easiest chemical test to try, and when Rutherford dipped a stick of bismuth into his solution of the active deposit, it came out with an alpha-particle activity. What was more, the penetrating power of its alpha rays exactly matched with a proper specimen of radiotellurium which he had bought. Very soon he discovered also that the alpha activity of his solution was slowly rising, which made it seem that the beta-ray substance it contained might be the parent of radiotellurium.

Another confusing material was radiolead. It came from pitchblende by the chemical processes which brought out lead, and the reports on its radioactivity were at least as confusing as those on polonium. Rutherford had a specimen prepared by Bertram B. Boltwood, a former Yale chemist who was now busy at research in his own private laboratory. When he dissolved it, and dipped in the bismuth stick, he drew out the alpha-ray material again, leaving behind something which gave out beta rays.

Along with these investigations, Rutherford was collecting information from the active deposits. His routine now was to expose a platinum strip (negatively charged) to the emanation for four days. After that, the beta-particle activity of the strip would climb in the familiar geometric style, every six days picking up half of what it lacked of a future, steady value. This meant that there was something on the platinum that did not give beta rays, and transmuted at a steady rate into a new substance that did. The new substance, which gave the beta rays, would have six days for the half-value period of its transmutation.

Eventually the strip would show an alpha-particle activity, and if he heated it now, the alpha particles would vanish, showing that the substance which gave them had evaporated. At the same

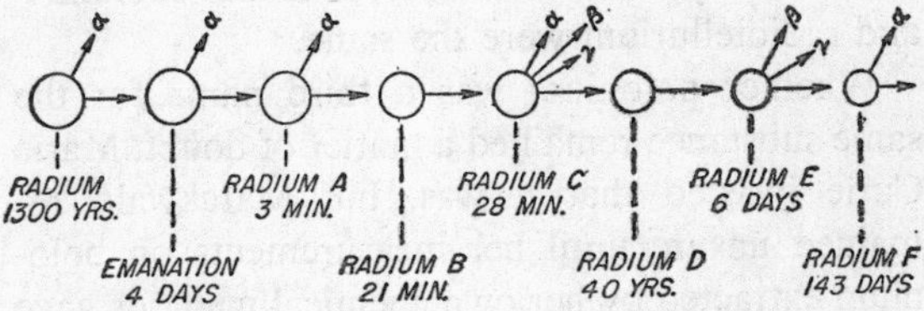

RADIUM 1300 YRS.   RADIUM A 3 MIN.   RADIUM C 28 MIN.   RADIUM E 6 DAYS

EMANATION 4 DAYS   RADIUM B 21 MIN.   RADIUM D 40 YRS.   RADIUM F 143 DAYS

Fig. 16. THE RADIUM CHAIN *of transmutations.*

time, the beta activity would stop growing and shift over to an ordinary, six-day, geometric decay, and that meant that its forerunner had evaporated also. If he waited now, the alpha particles would reappear as the beta particles from the strip grew fewer, and this made it plain that the six-day beta-giver was transmuting into the alpha-ray material.

To bring in the alphabetical names, it seemed that radium C must turn into a radium D which was rayless and had a rather long life. Radium D transformed into radium E which gave beta particles and had a half-value period of six days, and radium E into radium F, which gave alpha particles and had a half-value period of 143 days. (Fig. 16)

There was an able team of radioactive investigators at the University of Vienna, Stefan Meyer, a physicist, and Egon von Schweidler, a physical chemist. They had been studying radiotellurium, and they gave the half-value period of its radioactivity as 135 days. Marckwald, after a public debate with Soddy, had come to realize that no radioactivity could be really permanent, and he had set two young physicists named Heinrich Greinacher and Karl Herrmann to work on the radiotellurium he had prepared. They gave its half-value period as 139.8 days. It was certain then that radium F and radiotellurium were the same.

Whether polonium was a third name for the same substance remained a matter of doubt. Marie Curie insisted that it was, but Marckwald remained unsure until her measurements on polonium extracted by her own chemical methods gave a half-value period of 140 days. Then he gave in, and withdrew the name of radiotellurium. ". . . a rose by any other name would smell as sweet," he declared, and this Shakespearian surrender so delighted Rutherford that he boomed the quotation around the laboratory for a week.

As for Giesel's beta-ray polonium, it turned out to decay with a half-value period of six days, so it was the same as radium E. Radium D appeared to be the substance which was extracted as radiolead, and since the original radium D was always turning into radium E which turned rather quickly into radium F, the confusion of the earlier chemists was easily explained. The longer they waited between operations the more complicated their specimens grew.

By the spring of 1905, Rutherford had the chain

of transformations worked out, although the polonium controversy ran on into the summer of 1906. The transformation theory had done very well in settling the confusions of radiochemistry, but this was not the limit of its powers. It could suggest a plausible identity for the non-radioactive descendant of radium F.

Boltwood had been analyzing radioactive minerals with elaborate care, and the noticeable proportion of lead which they all contained struck him as interesting. It might be the accumulation of spent radium which had transmuted through emanation and all the alphabet from A down to F, and this, as Rutherford pointed out, was very plausible. Counting radium, the emanation, radium A, radium C, and radium F, five alpha particles came out along the chain. Since the experiment of Ramsay and Soddy, it had been reasonable to think that the alpha particles might be atoms of helium, and since helium had an atomic weight of 4, five alpha particles took away 20 units of atomic weight. Since radium had an atomic weight of 225, the element left after the transmutation of radium F would have an atomic weight of 205, and this was not far from the value of 206.7 which the chemists gave for the atomic weight of lead.

To think that the commonplace, dull lead of plumbers might have had a radioactive past was a fascinating idea.

11. The Descent of Radium

Radium had been found nowhere but in the minerals of uranium, and since its life was measured in geologically brief thousands of years, it must have been formed within the mineral and out of its substance. The only plausible material it could have come from was the uranium itself.

It was one thing to imagine this argument, and another to prove it. Soddy had begun confidently enough as soon as he reached Ramsay's laboratory in London. He dissolved a kilogram of uranyl nitrate in water, mixed in sulfuric acid, and drop by drop added a solution of barium nitrate. The barium sulfate which instantly precipitated could be expected to draw down any radium which might have been present, and for safety's sake he repeated the process several times over. Then he stoppered the bottle, waited a week for the emanation from any last residue of radium to accumulate, blew air through the solution to sweep out the accumulated emanation, and measured its radioactivity with a delicate gold-leaf electroscope. In another bottle he made up a solution with a carefully weighed morsel of pure radium bromide, blew air through it, and measured the emanation in the same way. Comparing results, he was sure that he had left no more than 10^{-11} grams of radium in the kilogram of uranyl nitrate, and, knowing this, he set the bottle aside for the uranium to turn into radium.

Unfortunately, by this time Ramsay was in full flight with his emanation experiments, and between the used emanation which he discharged into the air, and that which simply leaked out, the walls and ceiling of the laboratory and every piece of apparatus had picked up a coating of the active deposit. The electroscopes and electrometers discharged of their own accord, and delicate measurements became impossible. It was not until the fall of 1904, when he established himself in a clean laboratory of his own at the University of Glasgow, that Soddy could make any headway. There, to his disappointment he found that what little radium he could detect in his uranyl nitrate was less than the five-hundredth part of what he expected.

Now there was another way of testing to find whether radium was descended from uranium, and this was to discover whether the proportion of radium to uranium in different minerals was fixed or not. The argument on which it was based came from the geometric law of radioactive transmutation. This was generally given by saying that one-half of a particular element transmuted itself in so many minutes or hours or days. The half was only a convenient fraction, in fact the same kind of statement could be made using any other. (Radium emanation which was half-transformed every 4 days, was one-sixth transformed every 25 hours and one-tenth transformed in 14.5.) The law could also be stated by saying that during any particular length of time, a second or an hour or a year, some definite fraction of the element was always transformed. (For radium emanation, this would be 72 ten-thousandths every hour or 42 thousandths in a day.)

For uranium, over any reasonable time, this fraction was incredibly minute. For every practical purpose it was safe to say that from year to year or even from century to century, the number of uranium atoms transmuting themselves remained the same. As we have seen several times over, when there is a steady transmutation of one radioactive element, there is a gradual accumulation (up to a certain fixed value) of the next element in the chain. This fixed value would be reached when as many atoms of thorium X were turning into emanation as were being freshly formed from the thorium, or as many atoms of radium E became radium F as were being produced from radium D. Since the number passing on was always a fixed fraction of the atoms in existence, it followed that the quantity of the element which finally accumulated would be inversely proportional to the fraction of it which transformed.

All this was quite clear to Herbert N. McCoy, a young chemist at the University of Chicago (it must be admitted that the argument looks simpler in the calculus). He began the necessary measurements during the winter of 1903–4, while Soddy in London was preparing to grow the radium directly. They ran slowly, for he had to make sure that his ionization chambers measured what he wanted, and in the end that meant making allowances for the alpha particles which the sample itself absorbed. Then for each sample he must go through the tedious process of a quantitative analysis, carrying out each chemical operation so carefully and so deftly that the specimen which he weighed at the end would certainly contain every scrap of the uranium with which he had started and nothing else whatever.

When he had finished, in the spring of 1905, he could say with a good deal of authority that in the five different pitchblende ores he had studied the total radioactivity was exactly proportional to the amount of uranium they contained. That meant that every radioactive element in the ore, radium included, existed in a fixed proportion to the uranium, and so they all must be descended from it.

What McCoy had seen in Rutherford and Soddy's transformation theory, Boltwood of New Haven had also seen, but he managed to center his attack more directly on the radium. What he did was dissolve the mineral in acid, boil the solution, measure the radioactivity of the emanation carried off in the escaping steam, and so assess the amount of radium it contained. Like McCoy, he found the radium and uranium in absolutely fixed proportion, first for five different minerals, then for eight, and finally for sixteen, from deposits scattered widely over the face of the earth. At the same time he discovered that the proportion of thorium to uranium varied enormously between one mineral and another, so that these two elements must be radioactively independent.

It was clear then that radium was descended from uranium, but Soddy's failure to grow any made it clear also that it was not descended very directly. Some other element with a very long period of transmutation must fall between.

The idea interested Boltwood, and he thought first of actinium. This was another of the elements obtained from pitchblende, which had been discovered in the Joachimsthal residues by André Debierne in 1899 when he had first begun to work for the Curies. Boltwood had some of his own, which he had once extracted from a Colorado carnotite by Debierne's processes. He rummaged it out and was pleased to find that he could boil a recognizable radium emanation from its solution. He then purified it carefully, sealed it up in a glass vial, and set it away over the summer of 1906. Six months later he broke the seal, and now the emanation told him that 8.5×10^{-9} grams of radium had appeared there during the half-year of waiting.

This was too much success, for Boltwood had measured the total radioactivity of the uranium ores, and his own figures showed that actinium contributed very little to it, far less than the parent of radium would have to do. It was Rutherford who found the root of the trouble when he tried to repeat Boltwood's experiment for himself. From actinium there was a chain of transmutations through a radioactinium and an actinium X to a very short-lived emanation and an active deposit. (Fig. 17) By watching the growth and decay of activity in his different solutions, Rutherford found that the radium grew in a solution which contained no actinium but only radioactinium, and to make things worse, that the radium grew more slowly than the radioactinium was transmuting away.

Since the parent of radium was certainly where

the radium grew, its chemistry must be like the chemistry of radioactinium; and, acting on this hint, Boltwood ran it down early in 1907. It came as it should from uranium ores, processes which worked with thorium were efficient in extracting it, it had a high radioactivity of alpha particles, and it gave off ever-increasing amounts of radium emanation. Its alpha particles had a remarkably low

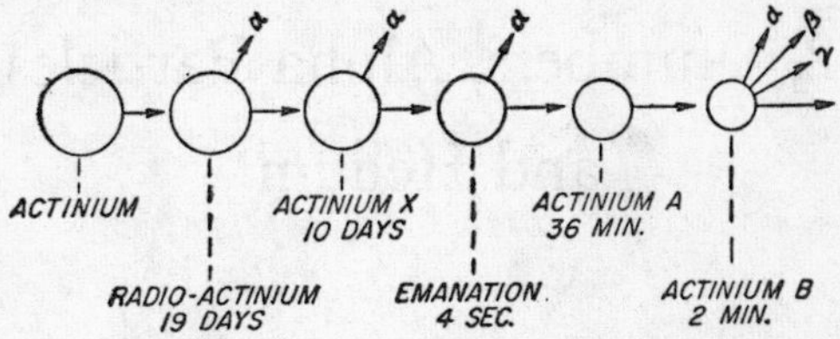

Fig. 17. THE ACTINIUM CHAIN

power of penetration, and in fact the whole combination of its chemical and radioactive behavior made it seem so different from everything else, that Boltwood felt safe in risking a name, calling it ionium and claiming it as a new element.

By the summer of 1908 it was quite clear that he was right. Marckwald and a student of his named Bruno Keetman confirmed the discovery, extracting what was plainly Boltwood's ionium from another uranium mineral. The only thing odd was its chemical behavior, for it gave an extraordinarily faithful imitation of the chemistry of thorium.

12. Numbers, Alpha Particles, and Helium

In 1905, while the experiments on radium D, E, and F were running out, Rutherford succeeded in measuring the current of positive electricity which a stream of alpha particles carried. Basically it was a simple experiment. He dissolved a little radium bromide in a good deal of water, he let some of this solution evaporate on a metal plate, depositing a coating of radium bromide which was thin enough to let most of the alpha particles through, and then he set another metal plate opposite to catch the alpha particles and register the charges they brought with them.

Practically, there were difficulties. The plates had to stand in a first-class vacuum, since any air between the two plates would contribute extra ions and confuse the measurements. Even then the alpha particles failed to bring in positive charges for what turned out to be a rather odd reason. Only half of the alpha particles shot away from their base plate (since the radium bromide gave them off evenly in every direction). The other half shot into it and dislodged, in a kind of atomic spatter, quantities of slow electrons. Once he knew this, Rutherford brought these electrons under control with a strong magnetic field which swerved them around, back into the plate from which they came.

Then the alpha particles arrived alone and could be measured. On the base plate lay 0.484 milligram of radium bromide, assayed not by weighing such a minute amount but by comparing its gamma rays with those from a larger and more manageable specimen. (The gamma rays were a very penetrating variety, and were given out in only a few of the radioactive transformations. They ionized only slightly and were rather hard to detect, but their penetrating power made them ideal for comparisons like this since it made no difference whether the specimen being worked with was sealed up inside a test tube or spread out in an open dish.) The alpha particles from the test specimen carried up charge at the rate of 9.8×10^{-13} coulomb per second. Suppose that each alpha particle had the standard ionic charge (which Thomson's latest measurement rated at 1.13×10^{-19} coulomb); then this electric current would lead to a final calculation of 6.2×10^{20} alpha particles shot off each second from a gram of pure radium. This is certainly a large number (10^{20} would be a hundred billion billion), but the number of atoms of radium it took to make up a gram was even larger. At this rate only 54 in each hundred thousand of them would transmute in any second, and the half-value period for radium would come out at 1280 years.

This was as good a calculation as anyone had

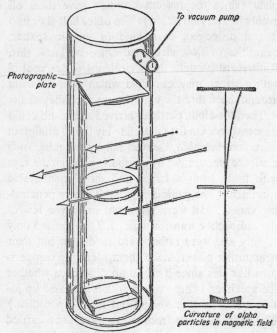

Fig. 18. MEASURING CHARGE-TO-MASS RATIO *of alpha particles. The alpha particles came from the radium C coating on a platinum wire placed in the V-block on the bottom of the evacuated container. They passed through the slit in the metal screen and through a magnetic field and registered on the photographic plate. The screen slit narrowed the beam of particles. The magnetic field, indicated by arrows, made the beam curve. From the location of the platinum wire, the screen slit, and the blackened line registered on the photographic plate, Rutherford could calculate the curve followed by the beam.*

made, but there was still too much guesswork in it. To do better, Rutherford would have to know something more about the alpha particles, so the next task would be to improve his measurement of their charge-to-mass ratio. With the radium he had now, the best way to get alpha particles was to expose a length of platinum wire to the emanation, let the radium A on it transmute away, and use the particles that the radium C gave out. If he mounted a wire like this below a slotted metal screen in a magnetic field, a photographic plate some distance above the slot showed a narrow line of blackening, and from the position of wire, slot, and line, it was easy to work out the curve the alpha particles had followed. (Fig. 18)

That was half the experiment, but the second half was harder, to get an electric field which would swerve the alpha particles without sparking over. Part of the solution lay in a really good vacuum; the rest came with ingenuity. The electric field Rutherford needed would be set up between two metal plates which were charged with opposite kinds of electricity. By making these plates rather long, so that the alpha particles traveled for some distance between them, he made sure that each alpha particle spent a reasonable time in crossing the field. Since it was being nudged sideways by the field for the whole period of its crossing, a gentle nudge from a moderate field might build up a fairly large deflection. As the apparatus developed, he also found it worthwhile to set the deflecting plates very close together. As experiments thirty years before had shown, in a good vacuum sparks require a considerable distance to develop, and simply will not pass over short gaps.

By the summer of 1906 he had finished. The charge-to-mass ratio came out at 5100 abcoulombs per gram. That made it about half of the charge-to-mass ratio for the simplest ion known, the hydrogen ion of electrolysis, and this was a hydrogen ion which had acquired a positive charge by giving

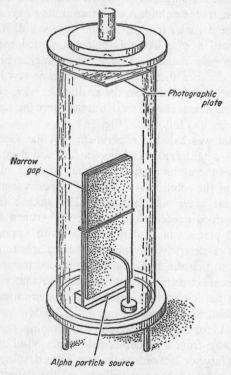

Photographic plate

Narrow gap

Alpha particle source

Fig. 19. SECOND STEP *in measuring charge-to-mass ratio of alpha particles. Alpha particles traveled through the narrow gap between two charged plates and on to the photographic plate. The electric field deflected the particles.*

up one electron. To halve any charge-to-mass ratio, you might halve the charge in the numerator or double the mass in the denominator, but for the charge-to-mass ratio of hydrogen, neither of these was particularly plausible. For the numerator there was no reason whatever to think that half-electrons existed. For the denominator, you could double the mass by linking together two hydrogen atoms to form a hydrogen molecule, but it was hard to see how a hydrogen molecule could survive the violence of a radioactive transmutation without breaking into its separate atoms. You could also chop a helium atom in two, but half-atoms were hardly more likely than half-electrons.

Anyone who makes change knows that two quarters are as good as one half. An alpha particle which carried twice the charge of a hydrogen ion and had four times its mass would come out with the proper charge-to-mass ratio. This would be an atom of helium which had ionized by losing two electrons, and it made of course a thoroughly reasonable guess.

Yet it seems to have tantalized Rutherford to be forced to guess. He already had half the information which would make him certain, for he knew the total charge which the alpha particles carried away from a gram of radium. All he lacked was a way of counting those alpha particles as they came off.

Three years before this, Crookes had been working with a screen coated with fluorescent zinc sulfide, which lighted up under the alpha rays, as Giesel had found, and could be bought ready-made from *Buchler und Compagnie*. By a slip of the hand, he had contaminated his screen with a speck of radium nitrate, and when he put it under the microscope to locate the intruder, he had been fascinated to discover that the light was not spread in an even glow over the screen, but came in a rapid series of separate, minute, and brilliant sparklings. Once he had seen it, a little experimenting brought the spectacle under his control. The radium need not be on the screen, but only near it, and a good hand magnifier was enough to show the separate sparks or scintillations.

By 1906, Rutherford knew the effect well. He and his students at McGill had used it steadily to detect alpha particles and to see how far they would travel. There was no doubt that each separate scintillation marked the arrival of a single particle, but there was no guarantee that every particle which reached the screen struck out its flash of light. No one really knew just how the light was produced, or whether the zinc sulfide might be sensitive all over, or only in particular spots.

What Rutherford did know was that a sensitive electrometer ought to give a perceptible swing of its needle if it was fed with all the ions which a single alpha particle could produce. So he turned to devices for ion-collecting, but the margin between what ought to occur and what actually did was a little too narrow. The electrometer was just not quite dependable.

In any case, the early months of 1907 brought an interruption. At the University of Manchester in England, Professor Arthur Schuster was anxious to retire, and quite as anxious to bring in Rutherford as his successor. Rutherford had done well at McGill. He had a secure position, and from the very start there had been plenty of time and money for his research. Nevertheless McGill was remote. The scientific center of the world still lay in Europe, and over the whole of the North American continent research in physics remained a luxury which even wealthy universities could hardly afford. To be nearer the heart of things, Rutherford decided to move, and by the middle of May, with the teaching year behind him, he took ship from Quebec for his new post in England.

In Manchester, he found already established a young German named Hans Geiger, fresh from his Ph.D. at the University of Erlangen and holding the appointment of John Harling Fellow in Physics. This meant that he was to be neither teacher nor student; his only business at the university was to do research. Rutherford always had projects, and Geiger was soon at work developing the ionization method into a practical scheme for counting alpha particles.

This turned out to be simply a matter of invention, which is not to say that it was easy, but only that the basis for the scheme was already known. It went back to a discovery by John S. Townsend, an old Cambridge friend of Rutherford, and to the days when Rutherford was new in Montreal. Just before this, when they had worked together on the X-ray ionization, Thomson and Rutherford had found it rather easy to collect all the ions in a space between two plates just as rapidly as the X-rays could produce them. Once this state had been reached, no increase in voltage on the collecting plates brought any more ions in, and they spoke of the ion current as being "saturated." Then Townsend had found that a large increase in collecting voltage could bring in more than the saturation quantity of ions.

What would happen was this. Under ordinary conditions, as an ion made its way to the collecting plate from the spot where it was produced, it would bump a good many times against neutral molecules of the air, but always moving so slowly that it simply glanced off and went on its way. Under Townsend's voltages, the electric field which moved the ions was strong enough to speed them up between collisions, so that every now and then the jolt with which an ion struck a molecule was enough to knock an electron from it. This made a new pair of ions to join the original set, and since these ions could also ionize the molecules into which they crashed, it was quite easy to build up the ionization by hundreds or even thousands of times.

(Here, by the way, you can see why it helped when Rutherford set the electric-field plates close together in the experiments on the charge-to-mass ratio of the alpha particle. In a good vacuum, there would be rather few gas molecules left in the narrow space between the plates. An ion which happened to be there could usually cross from one plate to the other without striking a single molecule, and so had no chance to increase the number of its companions enough to make a spark.)

Borrowing details from Townsend, or rather from a young mathematician named P. J. Kirkby, who had worked under him, Rutherford and Geiger took a long slender tube of brass, plugged its ends with hard-rubber stoppers, and strung a thin wire from end to end along its center. (Fig. 20) A hole through one stopper, covered with a thin plate of mica, let alpha particles into the tube, and a 1400-volt storage battery made the wire positive and the tube itself negative. Under these conditions, as Kirkby had shown, only the free electrons, driving inward toward the wire, could pick up enough energy to ionize when they collided. The positive ions would drift slowly and harmlessly

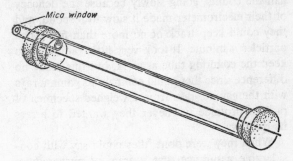

Fig. 20. THE SUCCESSFUL COUNTING TUBE. *Alpha particles entered the brass tube through the mica window. The tube itself was charged negatively; the wire running along its center from end to end was charged positively. An electrometer was connected to the wire. This was the original model of today's familiar Geiger counter.*

outward, while a steadily increasing cloud of electrons would rush toward the center together. When they reached the wire all at once, its charge would drop suddenly, an electrometer connected to it would acknowledge this drop by a swing of its needle, the battery would recharge the wire through a slow leak, and the tube would be ready for the next alpha particle.

The first tube built to these specifications counted nobly when alpha particles were let into it, and nearly as well when they were shut out. The fault seemed to lie in a "natural radioactivity" of the tube, and was cured very simply by making the next ones shorter and narrower and so reducing the amount of metal which might add its alpha particles to those sent in from outside.

When they got to a dependable shape, they began the counts, going slowly because the delicacy of their electrometer made it slow to respond, and they could keep track of no more than five alpha particles a minute. It took very little radium C to keep the counting tube at work, but this made no difference since they could compare its gamma rays with the gamma rays from a weighed specimen of radium bromide whenever they wanted to assess its quantity.

When they were done, they could say with considerable assurance that a gram of pure radium would shoot out 3.4×10^{10} alpha particles per second. Two years earlier, in Montreal, Rutherford had thought there were twice as many, but then he had thought that each alpha particle carried the charge of a single electron.

The counting had gone so well, that Rutherford thought it worthwhile to measure the alpha-particle current again, and once he had that figure, the simple division of the total number of alpha particles into the total charge they transported gave him the charge on a single one. It came out at 9.3×10^{-10} "electrostatic units."

The charge of one electron (and this was surely the unit of charge in the world of atoms) had been measured several times during five years before. J. J. Thomson had made it 3.4×10^{-10} (in electrostatic units), and H. A. Wilson, also at the Cavendish Laboratory, 3.1×10^{-10}. More recently R. A. Millikan and Louis Begeman of the University of Chicago had reported a figure of 4.06×10^{-10}. It was possible (though not likely) that the alpha particle carried a single electron-charge; it was plausible (and very likely) that it carried two, but it could not reasonably have any more. Rutherford and Geiger's figures made it necessary to allow the alpha particle its double charge and so fix it definitely as an atom of helium. They made it necessary also to raise the charge of the electron to 4.65×10^{-10} electrostatic units, which in modern terms would be 1.55×10^{-19} coulombs.

Fixing the electron-charge, and number of alpha particles given off by a gram of radium gave a new precision to all sorts of interesting calculations, of which we need notice only one. Rutherford and Geiger could say very definitely now that the half-value period for radium was 1760 years.

This is not quite the end of the story, however. By the summer of 1908, when Rutherford and Geiger had finished, it was plain to anyone who would follow the arguments that the alpha particles were atoms of helium. Three months later, although the matter was settled, Rutherford seized upon a chance to make it plain to anyone whatever.

By now, Rutherford had nearly a third of a gram of radium, loaned to him by the Academy of Sciences in Vienna, which had been working up the pitchblende residues at Joachimsthal on a large scale. He had a student named Thomas Royds, who had been trained in spectroscopy by Schuster. Also he had learned enough about both radium and the spectroscope to be sure that the experiment would work.

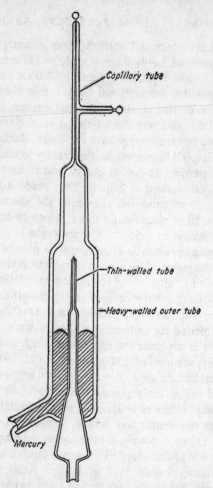

Fig. 21. APPARATUS FOR SPECTRUM EXPERIMENT.
The "emanation" was pumped into the thin-walled inner tube. Alpha particles given off by the "emanation" went through walls of small tube into the heavy-walled outer tube, from which the mercury forced them up into the capillary discharge tube at the top. It was fitted with electrodes for high voltage, and the spectrum was produced there.

He and Royds collected the emanation from a good quantity of the radium which he kept in solution. They pumped that emanation into a fine tube of very thin-walled glass, which it had cost the laboratory glass-blower, Mr. Baumbach, a good deal of trouble to make. (Fig. 21) As a scintillation screen showed, the alpha particles came through the glass very nicely. Then they shut the emanation tube inside a larger one of heavy glass, pumped the space between to a good, clean vacuum, and after a few hours, flooded that space with mercury to pump the accumulated alpha particles into a discharge tube at the top.

There was no helium to be seen in the spectrum, nor did any appear until two days had gone by. Then the first trace of the yellow line showed up and two days later the green one joined it, but it took nearly a week for the complete spectrum to develop. For all its slowness, that helium had still come through the glass as alpha particles, for when they pumped the emanation out and filled the fine-walled tube with helium in its place, no trace of it appeared in the spectrum tube, even after eight days.

As for the delay, the alpha particles which reached the outer glass wall of the apparatus must plunge into it as they had plunged through the glass of the inner tube. When they came to rest then, they would be deep inside the solid glass, and it would be no wonder if they needed a while to work their way out. If this was so, then a lead barrier should stop them sooner, and return them more quickly from the shallower trap it made. Rutherford and Royds tried the experiment again with a lead foil wrapped around the emanation tube, and this time the yellow and green lines appeared after one day and the whole spectrum after two.

Then for a final demonstration, they ran through it again with clean pieces of lead, and after a few hours melted the lead in a vacuum to see what gases they might release. In four hours a wrapping lead would gather enough helium to show a glimmer of the yellow line and in twenty-four the yellow and the green lines would be bright. Another piece from the same sheet, which had not been near the emanation tube, showed no helium whatever.

It was as dramatic an experiment as Ramsay and Soddy's had been. In fact it was the completion of that earlier experiment, for it made vividly clear how the helium into which the radium transmuted had come into being. Just as it finished, there came word from Stockholm that his work in transmutations had won for Rutherford the Nobel Prize in Chemistry for 1908.

13. Alpha-Particle Scattering

The science of radioactivity as we have met it so far was by no means worked out in 1908. There were still problems connected with the different transformation series, with the production of heat, with the distance an alpha particle would travel before it had spent its energy in ionization. It will be profitable for us, however, to turn aside to the matter of alpha-particle scattering, which occupied Hans Geiger off and on over the years between 1908 and 1912.

This was rather an insignificant effect which Rutherford had first noticed when he had been measuring the charge-to-mass ratio of the alpha particles with photographic plates at McGill. Unless he had a clean vacuum in the apparatus, there was never a sharp line where the alpha particles struck the plate, but only a blur, and even in a good vacuum the same blur showed up when he laid a bit of mica over the slotted screen which picked out the beam of alpha particles to be registered. It seemed that when they passed near (or perhaps through) the atoms of the air or mica along their path, the alpha particles were swerved a little from the line of flight they had been following. It was not a big swerve, no more than a fraction of a degree of slant, but when you began to reflect, as Rutherford did, on the enormous momentum of a flying alpha particle and the very small diameter of an atom, you found it hard to see how anything could be accomplished in so short a time or distance unless there lay inside the atom an electric field of thoroughly incredible intensity.

This "scattering" of alpha particles promised to tell something about the insides of atoms. The problem was to imagine a scheme for studying it, to think up an experiment to discover in what directions the alpha particles went, and how many in each direction, when they passed through a certain thickness of air or mica.

Rutherford had been in earnest when he set out with Geiger to count alpha particles. He needed the best information he could get, and he could not afford to tinker with methods which might not be reliable. Such demands did not hamper a young physicist named Erich Regener at the University of Berlin. When Professor Heinrich Rubens suggested that Regener should study the alpha-particle scintillations, he was quite willing to take them on, and to find out how reliable a detector the zinc-sulfide screen might be. He had a nice feeling for apparatus, and he worked out an admirable arrangement, neatly planned in every detail. His alpha particles came from a little polonium deposited on a copper strip, and fell on a layer of zinc sulfide painted on the front of a glass plate, while he watched the sparks of light with a microscope focused through the back. Then he counted, over and over, as long as his attention would hold steady, and concluded at last that his snippet of polonium must be giving out 1800 alpha particles per second. Then working back from the ionization current it produced (and assuming that each alpha particle carried twice the electron-charge) he got a second calculation of 2000 alpha particles per second, in very reasonable agreement.

This was in the early months of 1908 when Rutherford and Geiger's counting tubes had settled down to dependable work. As soon as they heard of Regener's success, they set up a scintillation screen for themselves, and, checking its sparkles against the swings of their electrometer needle, they found the same counts by either device.

The scintillation screen seemed just the tool they needed to study the scattering effect, and even while their last important counts were being made, Geiger ran off a test to see what scattering it might show. He had a thin, cone-shaped glass tube with a mica window sealed over one end, which he could fill with radium emanation to supply his alpha particles. This was mounted at one end of a long glass tube which he pumped down to the hardest possible vacuum. Well past the middle of the tube was a metal plate with a horizontal slit in it, and at the farther end, the zinc-sulfide-and-glass scintillation screen. Here, when the emanation tube was filled, there appeared a neat, sharp-edged, narrow bar of light, the silhouette in reverse of the slit-opening in the metal plate. Then if Geiger opened the tube, mounted a sheet of gold leaf over the slit, and pumped it down again, the band of light blurred out as the gold leaf scattered the alpha particles away from the straight lines of their flight. Behind the scintillation screen was a microscope, focused on a tiny portion of it, and mounted so that it could be slid up and down alongside a millimeter scale. Looking through this, Geiger could count the number of flashes per minute right at the center or near it or farther away, and get a fairly accurate description of how the scattering went. (Fig. 22)

The graphs he made by plotting up his counts showed very clearly the scattering was real. Although few of the particles were turned by as much

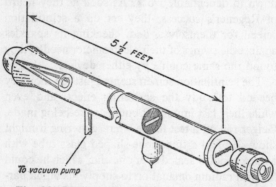

Fig. 22. GEIGER'S SCATTERING APPARATUS. *The cone-shaped glass tube at the left end of the device was filled with radium "emanation" and sealed with a mica window. The five-and-a-half-foot tube was evacuated. Alpha particles from the radium "emanation" traveled down the tube, through the slit in the metal plate to a zinc sulphide and glass scintillation screen at the right end. The alpha particles made a sharp, narrow bar of light on the screen, but when gold foil was placed over the slit, the band of light blurred. Through a microscope behind the scintillation screen, Geiger could count the flashes per minute and measure the scattering.*

as half a degree in the gold leaf, two-thirds of them were scattered to one side or the other, outside the rectangular silhouette of the slit. (Fig. 23)

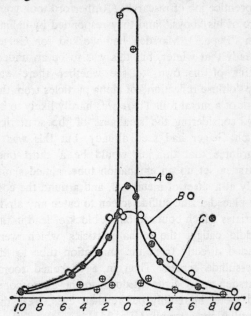

VERTICAL HEIGHT INDICATES THE NUMBER OF SCATTERED ALPHA PARTICLES PER MINUTE

Fig. 23. SCATTERING OF ALPHA PARTICLES. *Tests with the apparatus of Fig. 21 produced these curves. The horizontal scale represents the distance (in millimeters) of the scattering from the center of the light band on the scintillation screen. The A-curve (circles enclosing crosses) shows the behavior of particles when no foil was placed over the slit in the metal plate. The B-curve (white circles) shows the scattering with two gold leaves over the slit and the C-curve (black circles) with one gold leaf over the slit.*

Six months or so later, early in 1909, Rutherford asked Geiger to take on an undergraduate student named Ernest Marsden, and see him through an apprentice job of research. (Rutherford took good care of his "boys," and they responded by calling him "Papa.") Marsden had worked for Geiger already that winter, but this was to be an undertaking of his own, to see whether there was any "diffuse reflection" of alpha particles from the front of a metal foil. There was hardly likely to be any, considering the smallness of the scattering which Geiger had seen already, but this was a guarantee that the job would be a short one. Marsden set up an emanation tube, aimed slantingly at a sheet of metal foil, and around the corner placed a zinc-sulfide screen to catch any alpha particles which bounced off. A block of lead in the middle caught the alpha particles which were headed directly from the emanation tube to the zinc-sulfide screen. Then in a darkened room, Marsden and Geiger sat down to the microscope, and there, incredibly before their eyes, were the scintillations after all. (Fig. 24)

That made the project a good deal more interesting, and they began to explore. The alpha particles bounced back from foils of lead and gold, of platinum, tin, silver, copper, iron, and aluminum, and the higher the atomic weight of the metal in the foil, the more there were of them that bounced over. They tried piling sheets of gold leaf, one in front of the other, and found with the first half dozen that the number of scattered alpha particles grew directly with the thickness. That made it fairly certain that this was not a surface effect, that the alpha particles plunged in and were scattered within the body of the metal. Finally they made a count of the actual number of scattered particles, estimating from it that about one alpha particle in every eight thousand which fell on the foil was scattered back toward their scintillation screen.

So it happened that in the summer of 1909, when Ernest Marsden was graduated from the University of Manchester with the degree of B.Sc. and with First Class Honours in Physics, he found him-

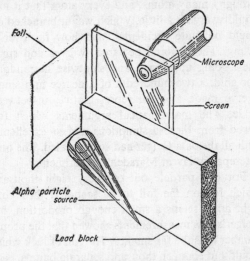

Fig. 24. THE GEIGER-MARSDEN EXPERIMENT *to investigate "diffuse reflection," or bouncing back, of alpha particles from a foil. The cone-shaped tube beamed the particles at the foil at an angle. The lead block shielded the scintillation screen from particles passing directly from the tube. Through the microscope the experimenters could see whether the foil did indeed "reflect" the particles to the screen.*

self also the junior author of a paper read before the Royal Society.

Another half-year went by, and in February 1910, Geiger sent in a paper of his own describing a long set of experiments on the alpha particles which were scattered in going straight through a metal foil. It was a detailed and careful piece of work, and what it seemed to say about the alpha particles was this. As each of them made its way through the metal, it would pass close to (or through) many atoms, and every atom that it met would swerve it a little. Which way it branched off would be quite accidental, and how far over it swung. There would be large swerves and small ones, going up, down, and slantwise at random. For gold, a two-hundredth of a degree in a single swerve made a good, representative, in-between value. The mathematical arguments which followed from these assumptions agreed excellently with all the data Geiger had accumulated. The only discrepancy lay in Marsden's experiment.

For one particle out of every eight thousand which fell on the foil to be scattered through a right angle seems a small enough proportion, but Geiger's new measurements showed that the proper number must be far smaller. The run of luck which would put eighteen thousand separate bumps, each of a two-hundredth of a degree, one behind the other in a single line, so as to run up a right-angled turn, was quite simply incredible. Yet Marsden's observations were true, and so were Geiger's, whether or not there was any way to reconcile them.

And so things lay from February to December; there were plenty of other promising investigations to keep people busy. Then Rutherford began to see how things might be explained, and soon he was proclaiming happily to everyone he met that now he knew what the atom looked like.

The key idea was this. The alpha particle found the foil almost literally empty. It was not likely that it would come near anything firm enough to turn it in its flight, and very unlikely in thin foils that it would do this more than once. Each large-angled turn that Marsden had seen had come from a single encounter between an alpha particle and some kind of scattering-center in the metal foil.

When Rutherford went on to imagine what kind of force an atom could exert, the most plausible kind was electrical. To turn an alpha particle abruptly would take an enormous electrical force, and this probably meant that the alpha particle came very close to the scattering-center as it swung around.

Suppose, to be specific, that an alpha particle from radium C was shooting with its proper speed of 2.09×10^9 centimeters per second, straight for the scattering-center of a gold atom. Suppose that this scattering-center carried a positive charge of a hundred electron-units, and suppose that Coulomb's Law might be used to calculate the electrical repulsion between the positive alpha particle and the scattering-center. Then it was easy to discover that the alpha particle would not be brought to a stop until it had pushed to a point only 3.4×10^{-12} centimeters from the scattering-center. The radius of an atom as it lay in a crystal or bumped against other atoms in a gas could be calculated as something like 10^{-8} centimeters. At the point where the alpha particle finally turned back, the edges of the atom were 3000 times farther away than the scattering-center was.

If the scattering-centers were as small as this, it would be no wonder if the alpha particles missed most of them entirely, or swung by at such a distance that their motion was very little affected. Yet for something close to a direct hit, such a scattering-center could swing an alpha particle drastically.

Coulomb's Law makes the force between two electrified bodies depend inversely upon the square of their separation. It is rather like the force of gravity in this, although electrical forces are much stronger than gravitational, and may be repulsions as well as attractions. Nevertheless, an alpha particle flying past a fixed scattering-center would behave remarkably like a swift comet near the sun. The laws of planetary motion would apply, and these laws required it to follow a hyperbola. The interesting thing was that the hyperbola it traced would have the same shape whether it swung around the back of a negatively charged, attracting scattering-center or was turned off in front of a positively charged, repelling one. The angle between the two arms of its orbit—that is to say, between the line along which it came in and the line along which it flew out again—would depend upon nothing but its aim, on how close to the scattering-center it would have passed if it kept on in a straight line. (Fig. 25)

As Rutherford realized this, he began to see how he could invent an experiment to test this atom-model. He could imagine target rings drawn around the scattering-center of each atom. If he considered all the rings of one particular radius, they would occupy a certain small area of the scattering foil. If he took the ratio of this area to the total area of the foil, then this ratio would be also the ratio between the number of alpha particles aiming

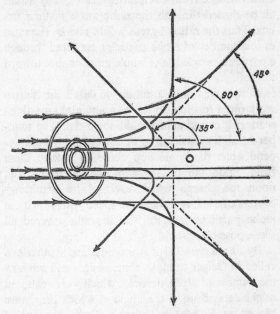

Fig. 25. ALPHA-PARTICLE SCATTERING. *The angle of scattering depends on how close the particle would have come to the positively charged scattering center (small circle) if the particle had not been turned aside. The large concentric rings at the left illustrate the scattering influence. Particles coming in on a line aimed at the outer ring scatter through an angle of 45 degrees. (Actually they would not reach the target.) Particles on the line aimed at the next ring toward the center scatter through 90 degrees, and on the line aimed at the central ring 135 degrees.*

toward one of those target rings and the whole number of alpha particles falling on the foil. Since these alpha particles all aimed themselves at the same distance from a scattering-center, they would all be turned through the same angle in their orbits. Thus the ratio of areas would also be the ratio of the number of alpha particles scattered through a particular angle to the whole number shot toward the foil.

Now it was only a matter of detail for him to work out a formula to tell how any alpha-particle-scattering experiment should come out. The number of alpha particles to be looked for would depend upon the angle you chose to receive them from, upon the thickness of the scattering foil, upon the charge which each of its scattering-centers carried, and upon the speed with which the alpha particles shot in. The formula covered all these possibilities.

By February 1911 it was ready, and Rutherford called in Geiger to make some counts and see how the number of alpha particles which were scattered might depend upon the angle at which they were picked up. By March, Geiger had some reasonable figures for "the Prof" (he was too German to use the nickname "Papa), and by April, Rutherford had written up all his ideas, describing the kind of atom which would scatter alpha particles and showing how the formula for the scattering could be derived. In May, with something like newspaper speed, *The Philosophical Magazine* published his paper—and then that was all. Geiger and Marsden picked up their old partnership and set about a careful, conscientious check of every detail of the scattering formula. Geiger's preliminary data were held back until the complete job would be finished, Rutherford turned to other business, and his atom-model rested quietly and almost unnoticed in the printed pages of one, purely theoretical paper.

14. Some Radiochemistry

It is high time now that we got back to thorium, and thorium brings us inevitably to Otto Hahn. He had begun as an organic chemist at Marburg in Germany, and, being offered a position if only he would learn English, he had turned up, to combine the English with other experiences, in Ramsay's laboratory in the fall of 1904. Ramsay had rather highhandedly set him to separate radium from a barium-containing fraction extracted from 250 kilograms of a new radioactive mineral from Ceylon. This required the kind of chemistry Hahn knew least, but he hunted out Giesel's papers and gave it a try. He found the radium soon enough, and an even more active substance with quite the opposite behavior, which did not crystallize with the bromides but hung back in the mother liquor. The new material gave an emanation, with the half-value period of one minute, and this laid down an active deposit with a half-value period of eleven hours. These should have been the signs of thorium, but no thorium was ever as vigorously ra-

142

dioactive as the stuff Hahn was handling. For the resemblances and for the one striking difference, Hahn settled on the name of radiothorium for what he felt sure must be a new, active substance.

Ramsay was mightily impressed, and promptly persuaded Professor Emil Fischer of the University of Berlin to engage Hahn as the radioactive expert of his research institute there. This was a bit rapid for Hahn, and in the summer of 1905 he sailed for Montreal to pick up a little experience in his suddenly acquired specialty. Once arrived, he was able to convince Rutherford that radiothorium was real, and that it, rather than thorium, was the predecessor of thorium X. Then he continued to work for a year on various physical and chemical problems connected with thorium and actinium, and finally left for his post in Berlin.

Radiothorium was real. Its alpha rays established it, and the way in which it would grow thorium X; but as a chemical element it soon began to make difficulties. It was descended from thorium as Boltwood and McCoy both made sure by showing that the total radioactivity of thorium minerals was strictly proportional to the thorium they contained. (McCoy was working now with a graduate student from Nova Scotia named W. H. Ross.) In Berlin, Hahn struck up an alliance with the Knöfler firm, the chemical manufacturer from whom Rutherford and Soddy had bought their purest thorium nitrate, and there he ran onto a very odd fact. Through the whole of their manufacturing process, from the crude ore to the most refined end-product, the proportion of radiothorium to thorium remained the same in every sample that he investigated. For some curious reason, the Knöfler process failed to remove radiothorium, although in Ramsay's laboratory it had been extracted with no trouble at all.

Then came a letter from Boltwood, with the casual news that he had a radiothorium preparation which had held a constant radioactivity over a period of two years. From his own measurements Hahn knew that two years was close to the time which radiothorium took to be half-transformed away. Boltwood's sample was certainly being replenished, and that meant that it probably contained a long-lived forerunner with a much longer life, which transmuted to radiothorium by a rayless change. Starting from this hint, he was presently able to extract from a mineral a substance which was not thorium and gave no rays, but which, as it stood, began slowly to develop an alpha-particle activity and to set free traces of the one-minute emanation. This was evidently the substance which came between thorium and radiothorium, and for this reason Hahn proposed to name it mesothorium.

Since Hahn had been given all kinds of help at the Knöfler plant, and since mesothorium might have a commercial future, he was careful to say nothing about its chemistry. Boltwood could take a hint as well as give one, however, and if radiothorium was rather like thorium, he thought that mesothorium might resemble thorium X. At this he hunted out an ancient specimen of thorium X, whose half-value period of four days should have left it long before as dead as a doornail. Now it gave off an unmistakable emanation, and the thorium X which released that must be freshly forming now from a radiothorium, which had come in its turn from some mesothorium in the original preparation. This made it seem, as Boltwood remarked, that Ramsay's barium fraction had contained mesothorium when it was new, and the radiothorium which Hahn had drawn from it so easily had probably grown there while it sat waiting on Ramsay's shelf.

Meanwhile, McCoy and Ross had been working diligently to separate radiothorium from thorium. They tried every trick of precipitation and crystallization, repeating some of their processes forty or a hundred times over. With perfect regularity they failed to make any change in the proportion of radiothorium to thorium in their working materials. At the end of 1907 they surrendered at last with the flat admission: "The direct separation of radiothorium from thorium by chemical processes is remarkably difficult, if not impossible."

Then Hahn turned up a beta-ray activity in his mesothorium, and by an elaborate use of the mathematical theory of transformations, he was able to work out the full chain of transmutation. Taking all the measurements made so far by Boltwood, by McCoy and Ross, by G. A. Blanc at Rome, and combining them with his own, he arrived at a thorium series which ran like this: Thorium gave alpha particles and transmuted steadily to the first mesothorium. This went by a rayless change into the second mesothorium and had a half-value period of five and a half years. The second mesothorium, with a half-value period of 6.20 days, gave out beta particles and transmuted into radiothorium. Radiothorium gave alpha particles, had a half-value period of nearly two years, and turned into thorium X. (Fig. 26)

By now, in the summer of 1908, the discovery of ionium had posed another "thorium" puzzle. Ionium descended from uranium and produced radium. It had nothing to do with the thorium series of transmutations. Yet, as Boltwood had discovered when he isolated it, whatever chemical reactions worked for thorium worked equally well for ionium. Hahn, in the same vein, found it in the purest of Knöfler's thorium nitrate, although the ore they had used was notably poor in uranium and would hardly have contained a whisper of

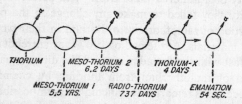

Fig. 26. THORIUM TRANSMUTATION CHAIN.

ionium. Then Keetman, Marckwald's student, backed up both Boltwood and Hahn. For a stringent chemical test, he had deliberately mixed ionium with thorium and tried to separate them again, and he had found no chemical trick which could even change the proportions of that mixture. Ionium was embarrassingly like thorium, and to this news Keetman added an extra note, that the processes which extracted ionium from the minerals also brought along uranium X.

You may begin to notice in here a certain monotony in the chemical behavior of these radioactive elements. Thorium and radiothorium and ionium and uranium X were all remarkably like one another, and the first mesothorium resembled thorium X. To this we can add another pair of details. In Marie Curie's laboratory at the Sorbonne (where she had succeeded to Pierre Curie's hard-won post) two separate attempts by chemists named Szilard and Herchfinkel failed to separate radium D from ordinary lead. Then in Berlin, in 1910, Marckwald discovered that radium contaminated with Hahn's first mesothorium was difficult if not impossible to purify.

Ever since he had left Ramsay's laboratory, in all the time between 1904 and 1910, Soddy had made miserable progress. His uranium solutions had steadily failed to grow radium, and when he had tried to work with uranium X instead, he had found it a thoroughly evasive material. Sometimes it would precipitate and sometimes it would not, and he could establish nothing definite in its chemical behavior.

Then in 1910 he wanted some mesothorium, and since Knöfler had not yet brought it on the market, he bought some thorianite and set out to extract his own. Remembering what Boltwood had said, he added barium to his solution (to duplicate Ramsay's operations) and brought the mesothorium down with a precipitate of barium sulfate. Then the mesothorium and the barium proved unexpectedly hard to separate, and he was driven to use the Curies' old process of fractional precipitation by alcohol. That worked nicely, the mesothorium chloride crystallized as he expected, but so did quantities of radium chloride, for his thorianite had contained nearly a third as much uranium as thorium. That was all right until he noticed that in every sample, and through every precipitation, the proportions of radium and mesothorium remained the same, and with them always came thorium X.

McCoy or Hahn or Marckwald might have met this calmly, but after the elusiveness of his uranium X, such forthright and definite behavior struck Soddy all of a heap. Where the others had spoken of chemistry which was "extremely similar" or "very much alike," Soddy declared baldly that in chemical behavior mesothorium 1, thorium X, and radium were "identical."

He knew quite well that he had no business to claim so much, and he met the difficulty head on. Since alpha particles were helium atoms, every alpha particle given out in a transmutation lowered the atomic weight of the substance by 4. Thorium had an atomic weight of 232.4, so mesothorium must have 228.4 and thorium X, 224.4. Radium, by direct measurement, had an atomic weight of 226.5. By the accepted rules of chemistry, three substances with three different atomic weights were three different chemical elements, and they could hardly have "identical" behavior. Soddy admitted all this, but rules or no rules, he had seen what he had seen, and he reinforced his argument by pointing to the other identity in the case of thorium, ionium, and radiothorium, whose atomic weights were 232.4, 230.5, and 228.4. Again, radiolead at 210.5 was identical with ordinary lead at 207.1. In the light of all this, he went on to suggest, such groups might be expected even among ordinary elements, showing the same whole-number differences in atomic weights, of which the world was ignorant only because the identity of their chemistry kept them always together.

It was a fascinating idea which drove him out of the laboratory and into "the literature" to look up all that had been written in radiochemistry. Through 1911 he searched and read, and put together his findings in a thin brown book called *The Chemistry of the Radio-Elements*. It was well worth doing, for out of this bookwork he pulled a very interesting discovery relating to a particular property of the Periodic Table.

When chemists talked of the "valence" of an element, they referred to the number of bonds which one of its atoms could extend to join onto other atoms in a compound. Oxygen was divalent, and each oxygen atom put out two bonds to link on two atoms of monovalent hydrogen in the molecule of water. Carbon was tetravalent, so an atom of carbon might link on four atoms of hydrogen in the molecule of methane, or two oxygen atoms, with a double bond apiece, in carbon dioxide. In the Periodic Table, all the elements in a vertical column had the same valence. In Column 1 and Column 7 lay elements whose valence was one, and the valence increased a unit at a time as you moved in from either side to reach the value of four in Column 4 at the center. (Fig. 27)

What Soddy noticed was that every radioactive element whose chemistry was understood had an even valence, so that it belonged in one of the even-numbered columns of the Table. The tetravalent trio of thorium-ionium-radiothorium belonged in Column 4 (under carbon), the divalent trio of radium-mesothorium-thorium X, in Column 2 (under beryllium). The inert emanations which had no valence took their place with the members of the argon family in Column 0. Polo-

O	I	II	III	IV	V	VI	VII	VIII
He 3.99	Li 6.94	Be 9.1	B 11.0	C 12.00	N 14.0	O 16.00	F 19.0	
Ne 20.2	Na 23.00	Mg 24.32	Al 27.1	Si 28.3	P 31.0	S 32.07	Cl 35.46	
A 39.9	K 39.10	Ca 40.1	Sc 44.1	Ti 48.1	V 51.0	Cr 52.0	Mn 54.9	Fe 55.8 Co 59.0 Ni 58.7
	Cu 63.6	Zn 65.4	Ga 69.9	Ge 72.5	As 75.0	Se 79.2	Br 79.92	
Kr 82.9	Rb 85.5	Sr 87.6	Y 89.0	Zr 90.6	Nb 94.	Mo 96.0	?	Ru 101.7 Rh 102.9 Pd 106.7
	Ag 107.88	Cd 112.4	In 115	Sn 118.0	Sb 120.2	Te 127.5	I 126.9	
X 130	Cs 133	Ba 137.4	La 139.0 / Yb 172.0	Ce 140.25	—	—	—	
	—	—	—		Ta 181.5	W 184.0	—	Os 191 Ir 193.4 Pt 195.2
	Au 197.2	Hg 200.6	Tl 204.0	Pb 207.1	Bi 209.0	—	—	
	—	Ra 226	—	Th 232.4	—	U 239.5	—	

nium, as Marckwald had placed it, lay below tellurium in Column 6, the lead into which it probably changed, in Column 4.

Whenever an atom gave out an alpha particle, the transmutation moved the element two places to the left in the Periodic Table, as when thorium changed to mesothorium 1, or thorium X to the emanation. Yet there were also backward movements, from mesothorium in Column 2 to radiothorium in Column 4, for example, or from radiolead in Column 4 up to polonium in Column 6, and then down (presumably) to Column 4 again.

Soddy had learned what was known, and he had also learned where chemical knowledge was lacking. In the early months of 1912, he enlisted a young Glasgow chemist named Alexander Fleck to work at filling the gaps, and by midsummer, Fleck was getting results. As Keetman had claimed already (although Soddy had found it impossible to believe), uranium X had the chemistry of thorium, and so, Fleck found, did radioactinium. Also thorium B had the chemistry of lead.

Fig. 27. THE *1912* PERIODIC TABLE. *This differs from the 1898 table (Fig. 1) only in the addition of a zero column for the inert gases, the insertion of radium in the last row, and changes in some atomic weights.*

15. The Radioactive Elements Find Their Homes

In 1912 there was a young lecturer in physical chemistry named Kasimir Fajans at the Institute of Technology in Karlsruhe in western Germany. Like Marie Curie, he had been born in Warsaw in Russian Poland, but instead of France he had chosen Germany for his education. After his Ph.D. at Heidelberg, he had gone to Zürich for a year of study and research, and then to Rutherford's laboratory in Manchester for another. Consequently he was full of knowledge of radioactivity, quite ready to take in everything Soddy had written in his book, and his fresh schooling in chemistry had given him a principle which promised to bring all the chains of radioactive transmutation into order.

Disregarding the inert gases, a full row of the Periodic Table started on the left with an alkali, which was "electropositive" since it went into solution as a positive ion, and ended with a halogen which went in as a negative ion and so was "electronegative." In between, the elements graded from one extreme quality to the other, so that you could always say that any particular element was more electronegative (or "electrochemically nobler") than its neighbor on the left. One useful test of this was the plating-out effect. Since polonium plated out of solution onto a stick of bismuth, polonium must be electrochemically nobler than bismuth, and so must go to the right of it in the Periodic Table.

Concerning some of the radioactive elements, as Soddy had already listed them, so much was already known about their chemical behavior that there was no question where they belonged in the Periodic Table. For others, it was known at least which of a pair was electrochemically nobler. Radium C for example was nobler than radium B, and radium F nobler than radium E which in turn was nobler than radium D.

This was the material that Fajans had to work with, and as he put it together, he saw how all the radioactive transmutations might be brought under two rules. When a beta particle was given out, the new element was electrochemically nobler than the old, or as he phrased it, the product was nobler than the mother substance. (A good many of the "rayless" changes had been discovered to occur with rather slow-speed beta particles so Fajans brought the rest of them under this rule also.) When an alpha particle was given out, the original element was electrochemically nobler than the new one. (Fig. 28)

Using these two rules, Fajans found it possible to fit every known radioelement into the Periodic

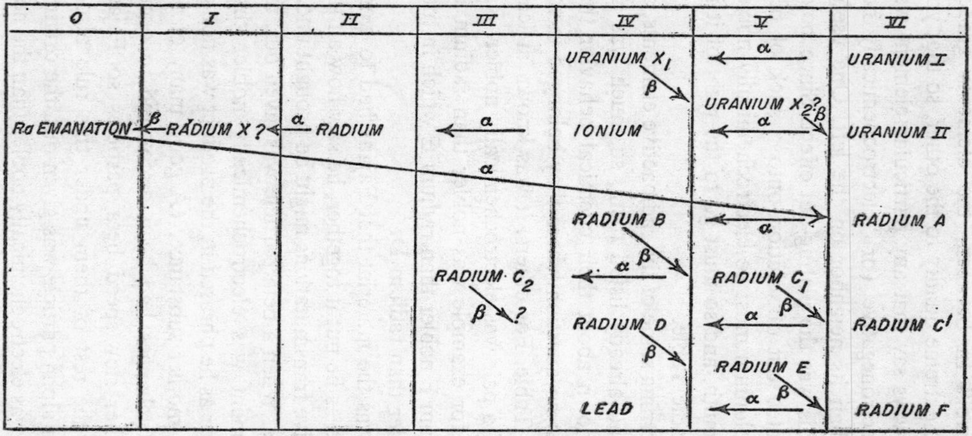

Fig. 28. URANIUM AND RADIUM TRANSMUTATIONS

	0	I	II	III	IV	V	VI
					URANIUM X_1	URANIUM X_2?	URANIUM I
							URANIUM II
					IONIUM		RADIUM A
				RADIUM C_2	RADIUM B	RADIUM C_1	
		RADIUM					RADIUM C'
	Ra EMANATION	RADIUM X ?			RADIUM D	RADIUM E	
					LEAD		RADIUM F

Table, if only, following Soddy's lead, he refused to worry about the number of elements of identical chemistry and different atomic weight which he had to cram into single places. Only at one or two points did he use his imagination to work around a difficulty.

Uranium X, in Column 4 (Fig. 29) gave a beta particle, and so must turn into an element in Column 5 or Column 6. Boltwood had discovered a few years before that uranium gave twice as many alpha particles as it should, so there could very well be two different "uraniums" in Column 6. The second uranium could not be the direct product of uranium X, however, for it gave alpha particles and the product of uranium X did not. Fajans had to suppose then that Crookes's uranium X, which he now called uranium X_1, transmuted to a second, very long-lived, uranium X_2, and this gave out slow beta particles to be transmuted into the second uranium.

The other problem involved the emanations. If they were produced by the alpha-particle transmutations of radium and of thorium X, then they must be less noble, and this meant that they were extremely electropositive. On the other hand, what little chemistry was known for the elements in the active deposits suggested that they were thoroughly electronegative, and could not be formed by an alpha-particle transmutation from an electropositive emanation. Fajans solved this by discovering an argument for classifying the emanations as extremely electronegative, and then inventing intermediate elements to make the transition. Radium, as he supposed, would give out its alpha particle and turn into a radium X in Column

Fig. 29. Fajans' Periodic Table

VI	V	IV	III	II	I	0
					Au 197.2	
				Hg 200.6		
			Tl 204.4			
		Pb 206.5	Ac D 206.5			
Ra F 210.5	Bi 208.4	Th D₂ 208.4	Th D 208.4			
Th C₂ 212.4	Ra E 210.5	Ra D 210.5	Ra C₂ 210.5			
Ra C' 214.5	Ac C 210.5	Ac B 210.5				
Ac A 214.5	Th C 212.4	Th B 212.4				
Th A 216.4	Ra C₁ 214.5	Ra B 214.5				
Ra A 218.5					Ac X₂ 218.5	Ac Em 218.5
					Th X₂ 220.4	Th Em 220.4
				Ac X 222.5	Ra X 222.5	Ra Em 222.5
				Th X 224.4		
		Ra Ac 226.5	Ac 226.5	Ra 226.5		
		Ra Th 218.4	Ms Th₂ 228.4	Ms Th₁ 228.4		
		IO 230.5				
		Th 232.4				
Ur II 234.5	Ur X₂ 234.5	Ur X 234.5				
Ur I 238.5						

1, and this would be very electropositive. Then radium X would give out a beta particle and so produce the electronegative emanation. (In the thorium series, a thorium X_2 would perform the same trick. Fig. 30)

As the elements went into their places, Fajans saw that he could add to his rules, and give not only the direction, but the distance an element moved in every transmutation. To give out an alpha particle shifted it two places farther down (as Soddy had already pointed out), to give out a beta particle, one place farther up. This fitted very neatly with the invention of uranium X_2, but the alpha-particle rule broke down in the case of thorium X_2 and radium X.

It took Fajans two long papers to argue this out, and they appeared in print in mid-February 1913, just as Soddy was finishing a manuscript of his own on precisely the same topic. The alpha-particle rule was his already, and Fleck's chemical explorations had shown him the way to the rule for beta particles. He and Fleck knew now that radium B had the chemistry of lead, and radium C and radium E both shared the chemistry of bismuth, and that placed these elements firmly just where Fajans had felt they ought to go. Soddy also invented a uranium X_2 (but called it eka-tantalum) to be set in Column 5. Since he depended upon chemical reactions rather than electrochemistry, he was spared the embarrassment the emanations had made for Fajans. Soddy's rules called simply for a two-place drop when an alpha particle came off, and so radium, in Column 2, produced the emanation in Column 0, and the emanation in Column 0 pro-

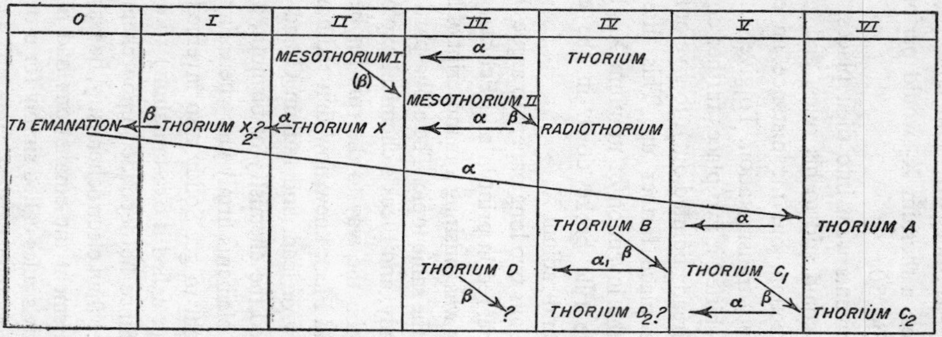

Fig. 30. THORIUM TRANSMUTATION

duced radium A in Column 6. Radium X and thorium X_2 were quietly left out.

Soddy had trifled with the rules already when he suggested that substances with different atomic weights and different kinds of radioactive behavior might still have only a single chemistry, and so must be taken chemically as varieties of a single element. Now he had another shock to deal out concerning Boltwood's ionium. It had been a matter of dispute what the half-value period for its transmutation might be, but now that eka-tantalum quite possibly existed, there were arguments to show that it might be as great as a hundred thousand years. In that case, ionium would be considerably more plentiful in minerals than radium.

The trouble was that ionium had no spectrum. An attempt had been made to photograph it at the University of Vienna by Franz Exner (an old friend to whom Röntgen had sent his pamphlet announcing the X-rays) and Eduard Haschek, two recognized spectroscopic experts, using a thorium-ionium sample which had been concentrated from ten tons of the Joachimsthal residues by the great Austrian chemist Auer von Welsbach. What they saw in the spectrum was thorium, then cerium and scandium in smaller quantities, and recognizable lines belonging to five other rare-earth elements, but no trace of anything which might have been ionium. Another attempt was carried out in Rutherford's laboratory at Manchester by Roberto Rossi (whom Schuster had trained in spectroscopy) and A. S. Russell (a former student of Soddy). Their thorium-ionium sample had been extracted by Boltwood, also from the Joachimsthal residues, during a winter's visit in Manchester, and after a second purification by Russell it showed the spectrum of thorium only, with the least trace of scandium.

Everyone agreed on the interpretation of these failures. There had been far less ionium in these samples than was expected, and that made ionium a transitory element with a rather short life. Now Soddy saw that the spectrum of ionium had actually been there. However upsetting it seemed, for all their difference in atomic weight, ionium and thorium must have identical spectra, just as they had identical chemical behavior.

Nevertheless by their radioactivity, it was still possible to tell the thorium called ionium from the thorium called uranium X, and these two from the thorium called radiothorium. It would be convenient to have a name for groups of substances with identical chemistry like this, which, although they were all one element, could still be told apart, and by and by Soddy invented the name "isotopes" from a pair of Greek words meaning "equal in place." It was distinctly useful since it was now much easier to state the question which Soddy had raised before (although at the moment there was no answer to it): Might not the ordinary elements also be made up of groups of isotopes of different atomic weights?

16. The Nuclear Atom

In the spring of 1912, when Soddy's book was only a few months old, a young Danish visitor named Niels Bohr had turned up at Rutherford's laboratory. He was one of the sons of the Professor of Physiology at Copenhagen, he had just taken his Ph.D. in physics there, and now he was finishing up a year of study abroad. Geiger and Marsden were still busy with the last long tests of the scattering formula, but Bohr found that in Manchester Rutherford's atom-model was already taken for granted. Everyone knew, that is to say, that the atom was as good as empty, that around its outer edge were electrons which moved in the electric field of a very tiny, very massive, positively charged scattering-center, and that this electric field fell off with the square of the distance outward. Bohr slipped easily into the life of the laboratory, talked, had ideas, made suggestions, listened with both ears, wrote a theoretical paper on the passage of alpha particles through matter, and after four months returned to Copenhagen. As his

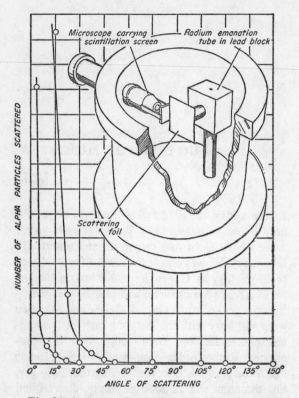

Fig. 31. ANGLE OF SCATTERING. *In this apparatus the lead block containing the radium "emanation" tube and the scattering foil were fixed to the base to keep the angle of aim constant. The box carrying the microscope and scintillation counter could be rotated on the base to change the line of sighting. The graph shows two scattering curves.*

162

visit came to its end, the positive scattering-center at the heart of the atom casually picked up a name, "the nucleus."

Two months later, Geiger returned to Germany after nearly six years in Manchester. The scattering experiments were finished at last; by October the paper describing them was ready. Then it waited, for the work had been done with the Viennese radium, and out of courtesy to the Academy of Sciences it was to be published simultaneously in Vienna and London. At last, in April 1913, the leisurely Austrian schedule permitted it to appear, and by then Bohr was well along on the task he had taken back home.

Rutherford had invented the nuclear atom only to account for the scattering of alpha particles (Fig. 31), and this was something it did superbly, as Geiger and Marsden had finally shown. Bohr wanted to make it more generally useful, to discover the way in which the electrons were held in place around the nucleus, and to see whether an atom built in this style might be handled so as to produce the spectra of the different elements and explain their chemistry and their radioactivity.

It was an impossible task from the start, for there was no proper way to build-in the electrons. If Bohr imagined them spotted in fixed positions around the nucleus, there could be no stability. If the repulsion of all their negative charges did not explode them outwards, the attraction of the positive nucleus must take over to collapse them inwards. If he thought of them as circling like the planets, the repulsions (which do not appear in the solar system) still came in to complicate the mathematics. Worse than this, if an electron circled, it had to accelerate; if a charged body accelerated, it radiated its energy in electromagnetic waves; if a circling electron lost energy, it must spiral down into the nucleus, and once more the atom would collapse.

How Bohr solved these problems is really not our business here. It required an extraordinary amount of inventiveness, to think up new rules which would be plausible and also work. As an extra difficulty, he undertook to build into his atom-model the new quantum theory of radiation which Max Planck had devised and Albert Einstein had been turning to good use. So through the fall of 1912 and the winter of 1913, Bohr tinkered with ideas, thinking his way through to the same useful answer first by one route and then by another, just to be sure that his arguments were reliable. At last in March, he had his first paper finished; in April it was ready for the editor, and in July it appeared. It dealt entirely (and successfully) with the hydrogen atom, which he took as one electron circling around a single, heavy nucleus.

His next paper, two months later, handled the questions which more complicated atoms brought up. A number of electrons would require, of course, an equal amount of positive charge on the nucleus to keep the atom neutral, but it was anybody's guess how many electrons there might be in an atom of nitrogen or tin or uranium. The most popular hypothesis was J. J. Thomson's, that the number of electrons equaled half the atomic weight, but Bohr decided on a different one. There lived in Holland a scientific amateur named Van den Broek, who had been educated as a lawyer and had fallen in love with the Periodic Table of the chemists. He was always looking for some simple, underlying rule which would explain everything in one swoop. Now Rutherford had once suggested, just to be logically complete, that a half-atom of helium would have the same charge-to-mass ratio as an alpha particle, and Van den Broek had seized upon it as the ideal building block for all the elements. Since it had a charge of one (on the scale of electrons) and an atomic weight of two, it gave directly Thomson's rule for the number of electrons in an atom, and besides, as a little clever arithmetic would show, on the average all atomic weights were even.

His arguments were fantastic, but one part of the idea seemed promising, and Bohr proposed to give each atom its quota of electrons by numbering the elements in order as they lay in the Periodic Table, and letting the order-number of each element be the number of electrons in its atom. That would give one electron to hydrogen, two to helium (so an alpha particle would be a bare nucleus), three to lithium, four to beryllium, and so on down the line.

The nucleus carried nearly all the mass of the atom, but it was unimaginably small, and that made it unimaginably remote from the atom's outermost parts. As Bohr pointed out, it would be these outer parts which gave an atom its chemical behavior or the particular pattern of its spectrum. Thus these qualities of an atom would depend only upon the arrangement of electrons around its outer boundaries, and this arrangement would depend chiefly on the number of electrons in the atom all told. Yet atoms which looked alike chemically, which had equal numbers of electrons, might still have nuclei with different masses and different kinds of radioactivity even though they carried equal positive charges. The nuclear atom accounted very neatly for isotopes.

The same argument led even further. If a nucleus lost an alpha particle in its radioactivity, the alpha particle carried off with it two units of positive charge. Two units less of charge on the new nucleus, and two less electrons moving about it, located the new atom two places down the Periodic Table, just where Soddy's rule required it to be. If a nucleus lost a beta particle—that is to say, a negative electron—it would gain a unit of positive charge, and so the new element must lie one place up the Periodic Table, just as Fajans and Soddy said that it would.

The new atom, which Rutherford had begun and Bohr was improving, seemed full of promise in 1913. It accounted for the spectrum of hydrogen, and might be expected to do as much for the other elements when the details of the electron arrangements could be worked out. It had shown a reason for that complete separation between ordinary chemistry and radioactivity on which Rutherford and Soddy had insisted when they first proposed their theory of transmutations in 1902. It had located the radioactivity of atoms in their nuclei, and it was clear now that tiny and remote as they were, those nuclei must be explored if radioactivity was ever to be understood.

This is the place then to break off our story. Not all the problems of radioactivity were solved yet, but the excitement would be passing from the investigations you have met here to others of a different kind. The future would lie with nuclear physics, but there is, nevertheless, one more chapter to add.

17. A Postscript on Lead

In 1913, radioactivity and chemistry were still worlds apart. To the ordinary chemist who had not followed the intricate linking of argument with experiment, the transformation theory seemed fanciful and imaginative, far too speculative to be believed. Yet now the most "scientific" chemists of all, who worked with numbers of the highest precision, the experts on atomic weights, were to give that theory another dramatic confirmation.

In the chains of transmutations which began with uranium and thorium, the last radioactive elements were radium F (or polonium) and thorium D. The new rules of Fajans and of Soddy required that the radium G and thorium E into which they transmuted must both be isotopes of lead. What was more, the subtraction of alpha-particle weights was enough to show that radium G would be a lead with an atomic weight of 206 (or a bit under), and thorium E, a lead whose atomic weight was 208 (and something over). The atomic weight of ordinary lead was just over 207, and halfway in between.

Radium G and thorium E had been accumulating in minerals at least since the time when those minerals were laid away in the rocks, and should be there in fairly sizable amounts by now. Then if you could find a uranium mineral without much thorium, and a thorium mineral without much uranium, and if neither of these was much contaminated by ordinary lead, you might be able with only ordinary chemistry to come out with some quite extraordinary atomic weights.

Since the argument was as easy as this, it is not surprising that a good many people started at once. In Glasgow, Soddy began, with an assistant named Hyman, to purify the lead in a thorite from Ceylon. In Paris, Maurice Curie, son of Pierre Curie's brother Jacques, started off with a handful of different minerals. In Karlsruhe, Fajans was more cautious. He knew the exacting work which atomic-weight measurements called for, and, feeling that this was best left to experts, he sent one of his students named Max Lembert across the Atlantic to the Harvard laboratory of T. W. Richards, bringing with him two samples of crude lead, one from a Colorado carnotite, the other from the Joachimsthal pitchblende. Finally in Prague, which was then a city of Austria, Otto Hönigschmid, a chemist who had studied with Moissan and with Richards, set out with an assistant named Stefanie Horovitz to investigate lead from the Joachimsthal residues.

By the summer of 1914, they were all ready to report. In preliminary experiments, with far too little material for accuracy, Soddy and Hyman had atomic weights of 208.3 and 208.5 for the lead from thorium. For lead from the uranium ores, there came a variety of values. Richards and Lembert had a high of 206.86 for lead from an English pitchblende sent them by Ramsay, Hönigschmid and Horovitz came in the middle with 206.736, Maurice Curie had a low of 206.36, and there were five others in between.

To an experienced radiochemist, all this was reasonable enough. None of the uranium specimens had contained pure radium G, but in every case enough of that new-born lead had been mixed in to pull the atomic weight of the whole mass definitely below the ordinary value. To Richards, however, it was completely amazing. He had made skeptics of the Harvard chemists. They refused to assume that the atomic weight of silver or sodium or iron was always the same, but insisted on testing samples of these elements from the four corners of the earth. Up until now they had always found as an experimental fact that wherever the raw material had been gathered the atomic weight of an element never varied from a single, identical value. Now, however, Richards had seen eight trustworthy and different atomic weights for eight different specimens, each of them beyond any doubt lead of the most exquisite purity, and he set out with tremendous vigor to discover whether ordinary lead and radium G might differ in any other property beside their weight.

By the end of 1915, he had a lead from an Australian ore which was definitely less dense than ordinary lead. By the middle of 1916, he had lead from some crystals of a very pure cleveite, sent from Norway by a radiochemist named Ellen

Gleditsch, who had worked with Boltwood in New Haven only a few years before. It was even less dense than the Australian sample, and when, by the fall of 1916, he had determined their atomic weights, the Australian lead gave 206.34 and the Norwegian 206.08.

So much for the difference, but the likenesses were quite as interesting. The densities of ordinary lead and of the Australian and Norwegian samples turned out to be very strictly proportional to their atomic weights. That meant that the atoms of radium G and of ordinary lead had very exactly the same volume. They had also, as he discovered over the next two years, the same spectrum, the same solubility in water, and the crystals of their nitrates had the same index of refraction. They were quite as identical as Soddy had thought isotopes ought to be, and as Bohr's new atom suggested that they must. The theories of radioactivity and the new atomic physics it was bringing into existence could hardly have had a more demanding test.

Meanwhile, in 1915, Hönigschmid and Horovitz had found their pure radium G. They had a uraninite from a place called Morogoro, in what was then German East Africa, whose lead had an atomic weight of 206.04. From Norway they had a bröggerite which gave 206.06. There could be no doubt whatever that uranium eventually transmuted itself into a light variety of lead.

What thorium became was a much more difficult question. Except for Soddy and Hyman's preliminary work, most of the lead from thorium minerals showed an atomic weight which was actually less than that of ordinary lead. This was because most thorium minerals contained large amounts of uranium, and because uranium transmuted itself three or four times more rapidly than thorium did. Nevertheless, it led to the suspicion that thorium E might have a slow, residual radioactivity, and the end of the thorium chain might then be an isotope of thallium or bismuth. Yet thallium and bismuth were notably scarce in most minerals of thorium. Then Soddy extracted a quantity of lead from the Ceylon thorite and sent it to Hönigschmid, who announced in 1916 that it had an atomic weight of 207.77. (At this time, Soddy was in Aberdeen and Hönigschmid in Prague. It is interesting that scientific communication could go on between them while Great Britain and Austria were at war.) Fajans followed with lead from a very pure Norwegian thorite, and for this Hönigschmid found 207.90. At last thorium E could take its place beside radium G. Each had its private atomic weight, and though neither was the common plumber's material, chemically speaking, neither could be called anything but lead.

18. The End

Isotopes and the nuclear atom, penetrating alpha and beta rays which were really streams of atomic particles, transmutations by the dozen wandering back and forth unhindered across ten places of the Periodic Table, all these are very different from Becquerel's first dream of making X-rays with ordinary light. No one could have foreseen what would follow the gray stain which slowly deepened on his photographic plate. Yet from that plate all the way to the atomic weight of thorium E, you have been walking a continuous trail. Each new idea has been the consequence of an earlier one, each fresh experiment suggested by something already known.

The advances of science are continuous, but it needs to be said again that they are not predictable. The alpha rays seemed unimportant because of their small penetrating power, yet they provided dramatic evidence for transmutation, made it possible to predict the chemical nature of radium G, and led Rutherford deep into the heart of the atom.

Radium was the exciting new element, but it was commonplace thorium which gave the clue to the transformation theory. McCoy and Boltwood and Hahn and Marckwald were gifted chemists who accomplished great things, and it was Soddy's failures which brought him to see isotopes in all the likenesses they had pinned down.

There is something else to consider too. You were promised at the start that this would be a story of atomic physics, and very little has been said about atomic physics anywhere along the way. This is as it should be. Atomic physics, as a special branch of science, would begin where we have ended, with the first successes of Rutherford and Bohr's atomic model. What we have been following here was all preparation for that.

It is easy to talk about dogs, for you have seen dogs and touched dogs, played with dogs and perhaps fed them or run from them. You have all sorts of experience to pack behind the three letters d-o-g. If you would rather talk about such fascinating beasts as a coelacanth or a pangolin, you may find that once you have learned to spell them, there is very little left that you can say. To go on, you would have to discover whether they lived in trees or burrows, whether they wore feathers or fur, whether they shed their horns or kept them. You would have to build up a body of experience about them.

This is the way it was with atoms. At the time this story began, "an atom" was a vague idea, as shadowy and undefined as perhaps these strange beasts are to you now. It was only very gradually that the idea of atoms became useful. J. J. Thomson's theory of the ionization of gases was easy to state in atomic language. Rutherford and Soddy needed atoms to describe transmutations, and when they imagined an alpha or beta particle shooting out at the instant the atom transformed, the transmutation itself became an atomic act, as separate and isolated in time as the atoms themselves would be thought to be separate and isolated in space. Rutherford was numbering atoms even before he was able to count them, but his electrometer swings and the flashes on his scintillation screen could hardly be talked of in any other terms. There was no one place where we could say, "Look! here the atom became real." It was rather that atoms grew real as they were used to describe the results of one experiment after another.

Do not think for a moment, though, that now you know the "real" atom. The atom is an idea, a theory, a hypothesis; it is whatever you need to account for the facts of experience. A good deal has happened since the closing point of this story, and the atom has been changing to keep up. A good deal will happen in the future, and the changes in the atom will continue. An idea in science, remember, lasts only as long as it is useful.

WHO'S WHO IN THIS BOOK

BARNES, HOWARD T. (1873–1950): Associate Professor of Physics at McGill University, Montreal, succeeded Rutherford as Professor of Physics there.

BARTHÉLEMY: French physician.

BECQUEREL, HENRI (1852–1908): Professor of Physics at the Museum of Natural History and at the *École Polytechnique* in Paris. Nobel Prize in Physics, 1903, shared with Curies.

BEGEMAN, LOUIS: Graduate student at the University of Chicago, later Professor of Physics and Chemistry at Iowa State Teachers College, Cedar Falls.

BÉMONT, GUSTAVE: Laboratory Instructor in Chemistry at the Municipal School for Industrial Physics and Chemistry, Paris.

BLANC, GIAN ALBERTO: Lecturer in Physics, later Professor of Geochemistry at the University of Rome.

BOHR, NIELS (1885–   ): Danish physicist, later Professor of Theoretical Physics at the University of Copenhagen. Nobel Prize in Physics, 1922.

BOLTWOOD, BERTRAM B. (1870–1927): Research chemist in New Haven, Conn. Assistant Professor of Physics, then Professor of Radiochemistry at Yale University.

CALLENDAR, HUGH LONGBOURNE (1863–1930): Professor of Physics at McGill University,

Montreal, later at the Imperial College of Science and Technology, London.

CROOKES, WILLIAM (1832–1919): Consulting chemist in London, owner and editor of *The Chemical News,* knighted in 1897.

CURIE, JACQUES (1856–1941): Lecturer in Mineralogy and later Professor at the University of Montpellier, France. Brother of Pierre Curie and father of Maurice Curie.

CURIE, MARIE (1867–1934): Wife of Pierre Curie, succeeded him as Professor of Physics at the University of Paris. Nobel Prize in Physics, 1903, shared with Becquerel, and in Chemistry, 1911.

CURIE, MAURICE (1888–   ): Student of radioactivity in Paris, later Professor of Laboratory Physics at the University of Paris. Son of Jacques Curie.

CURIE, PIERRE (1859–1906): Professor of Physics at the Municipal School for Industrial Physics and Chemistry, Paris, later Professor of Physics at the University of Paris. Nobel Prize in Physics, 1903, shared with Becquerel.

DEBIERNE, ANDRÉ (1874–   ): Laboratory Instructor in Chemistry, later Professor of Chemistry at the Municipal School for Industrial Physics and Chemistry, Paris, later at the University of Paris.

DEMARÇAY, EUGÈNE (1852–1904): Research chemist who maintained a private laboratory in Paris.

DESLANDRES, HENRI (1853–1948): Astronomer at the Astrophysical Observatory at Meudon, near Paris.

DEWAR, JAMES (1842–1923): Professor of Chemistry at the Royal Institution, London, knighted in 1904.

DORN, ERNST (1848–1916): Professor of Physics at the University of Halle, Germany.

EINSTEIN, ALBERT (1879–1955): Examiner in the Swiss Patent Office at Bern, later Professor of Physics at the Universities of Zürich, Prague, and Berlin, and at the Institute for Advanced Study, Princeton, N.J. Nobel Prize in Physics, 1921.

EXNER, FRANZ (1849–1926): Professor of Physics at the University of Vienna.

FAJANS, KASIMIR (1887–   ): Lecturer in Physical Chemistry at the Institute of Technology, Karlsruhe, Germany, later Professor of Physics and Chemistry at the University of Munich, then Professor of Chemistry at the University of Michigan.

FISCHER, EMIL (1852–1919): Professor of Chemistry at the University of Berlin. Nobel Prize in Chemistry, 1902.

FLECK, ALEXANDER (1889–   ): Student in chemistry at the University of Glasgow, now Chairman of Imperial Chemical Industries, knighted in 1955.

FRANKLAND, EDWARD (1825–1899): Professor at the Royal Institution and Professor of Chemistry at the School of Mines, London.

GEIGER, HANS (1882–1945): Research physicist at the University of Manchester, later at the *Physikalische Technische Reichsanstalt* in Berlin, still later Professor of Physics at the Universities of Kiel and Tübingen.

GIESEL, FRIEDRICH (1852–1927): Research chemist with the manufacturing firm of *Buchler und Compagnie,* Braunschweig, Germany.

GLEDITSCH, ELLEN (1879–   ): Assistant, later Professor of Chemistry at the University of Oslo.

GREINACHER, HEINRICH (1880–   ): Assistant in Physics at the University of Berlin, later Professor of Physics at the University of Bern.

GRIER, ARTHUR GORDON: Demonstrator in Physics at McGill University, Montreal.

HAHN, OTTO (1879–   ): Lecturer in Chemistry at the University of Berlin, later Director of the Kaiser Wilhelm Institute for Chemistry in Berlin. Nobel Prize in Chemistry, 1944.

HASCHEK, EDUARD (1875–1947): Lecturer, later Professor of Physics at the University of Vienna.

HERCHFINKEL, H.: Research chemist at the University of Paris.

HERRMANN, KARL (1882– ?   ): Student at the University of Berlin, later Associate Professor of Physical Chemistry at the Institute of Technology in Berlin.

HÖNIGSCHMID, OTTO (1878–1945): Research Associate in Chemistry at the German University in Prague, later Professor of Chemistry at Munich.

HOROVITZ, STEFANIE: Chemist at the German University in Prague.

HYMAN, H.: Chemist at the University of Glasgow.

JANSSEN, PIERRE JULES CÉSAR (1824–1907): Director of the Astrophysical Observatory at Meudon, near Paris.

KEETMAN, BRUNO (1884–1918): Graduate student in chemistry at the University of Berlin, later Director of radioactivity laboratory at Auer Gesellschaft, Berlin.

KIRKBY, PAUL J. (1869– ?): Mathematician at Oxford University.

LABORDE, ALBERT: Physicist at the Municipal School for Industrial Physics and Chemistry, Paris.

LECOQ DE BOISBAUDRAN, PAUL EMILE (1838–1912): French chemist who maintained a private laboratory in Cognac and later in Paris and preferred to be called François.

LEMBERT, MAX (? –1925): Student of chemistry, later Associate Professor of General and Inorganic Chemistry at the Institute of Technology in Karlsruhe, Germany.

LIPPMANN, GABRIEL (1845–1921): Professor of Physics at the University of Paris. Nobel Prize in Physics, 1908.

LOCKYER, NORMAN (1836–1920): English astronomer, later Director of the Solar Physics Observatory at South Kensington, knighted in 1897.

McCOY, HERBERT N. (1870–1945): Associate Professor, later Professor of Chemistry at the University of Chicago.

MACDONALD, WILLIAM C. (1831–1917): Tobacco manufacturer of Montreal who gave McGill University its scientific and engineering laboratories, equipped them, and endowed professorships for them, knighted in 1898.

MARCKWALD, WILLY (1864– ?): Professor of Chemistry at the University of Berlin.

MARSDEN, ERNEST (1889–   ): Student at the University of Manchester, later Professor of Physics at Victoria University College, Wellington, New Zealand, then Secretary of the Department of Industrial and Scientific Research, New Zealand, knighted in 1958.

MENDELEYEV, DMITRI IVANOVICH (1834–1907): Professor of General Chemistry at the University of St. Petersburg in Russia.

MEYER, STEFAN (1872–1949): Assistant in Physics, later Professor of Physics at the University of Vienna.

MILLIKAN, ROBERT A. (1868–1953): Associate Professor of Physics at the University of Chicago, later professor there, then Professor of Physics and Chairman of the Executive Council at the California Institute of Technology. Nobel Prize in Physics, 1923.

MOISSAN, HENRI (1852–1907): Professor at the Advanced School of Pharmacy, Paris. Nobel Prize in Chemistry, 1906.

NILSON, LARS FREDRIK (1840–1899): Professor of Analytical Chemistry at the University of Upsala in Sweden.

OUDIN: French physician.

OWENS, ROBERT BOWIE (1870–1940): Professor of Electrical Engineering at McGill University, Montreal, later Secretary of the Franklin Institute, Philadelphia.

PLANCK, MAX (1858–1947): Professor of Physics at the University of Berlin. Nobel Prize in Physics, 1918.

POINCARÉ, HENRI (1854–1912): Lecturer at the

École Polytechnique and at the University of Paris, later Professor of Mathematics and Astronomy at the University.

RAMSAY, WILLIAM (1852–1916): Professor of Chemistry at University College, London, knighted in 1902. Nobel Prize in Chemistry, 1904.

RAYLEIGH (JOHN WILLIAM STRUTT, LORD RAYLEIGH) (1842–1919): Research physicist, at one time Professor of Experimental Physics at Cambridge University. Nobel Prize in Physics, 1904.

REGENER, ERICH (1881–1955): Research physicist at the University of Berlin, then Professor of Physics at the Agricultural Institute in Berlin, later at the Institute of Technology in Stuttgart, Germany.

RICHARDS, THEODORE WILLIAM (1868–1928): Professor of Chemistry at Harvard University. Nobel Prize in Chemistry, 1914.

RÖNTGEN, WILHELM CONRAD (1845–1923): Professor of Physics at the University of Würzburg, Germany, later at the University of Munich. Nobel Prize in Physics, 1901.

ROSS, WILLIAM HORACE (1875–1947): Graduate student in chemistry, University of Chicago, later an agricultural chemist for the State of Arizona and the United States Department of Agriculture.

ROSSI, ROBERTO: Physicist at the University of Manchester.

ROYDS, THOMAS (1884–1955): Research student in physics at the University of Manchester, later

Director of the Kodaikanal Observatory, Madras, India.

RUBENS, HEINRICH (1865–1922): Professor of Physics at the University of Berlin.

RUSSELL, ALEXANDER SMITH (1868–   ): Research chemist at Glasgow, Berlin, Manchester, later Lecturer in Physical Chemistry at the University of Sheffield, then Reader in Chemistry at Oxford University.

RUTHERFORD, ERNEST (1871–1937): Research student at Cambridge University, Professor of Physics at McGill University, Montreal, the University of Manchester, and Cambridge University, knighted in 1914, made Lord Rutherford in 1931. Nobel Prize in Chemistry, 1908.

SCHUSTER, ARTHUR (1851–1934): Professor of Physics at the University of Manchester, later Secretary of the Royal Society, knighted in 1920.

SCHWEIDLER, EGON RITTER VON, (1873–1948): Lecturer in Physical Chemistry, later Professor, at the University of Vienna.

SODDY, FREDERICK (1877–1956): Demonstrator in Chemistry at McGill University, Montreal, then Lecturer in Physical Chemistry at the University of Glasgow, later Professor of Chemistry at Aberdeen University and at Oxford University. Nobel Prize in Chemistry, 1921.

SZILARD, BELA: Chemist at the University of Paris.

THOMSON, J. J. (1856–1940): Professor of Physics at Cambridge University, knighted in 1908. Nobel Prize in Physics, 1906.

TOWNSEND, JOHN S. (1868–1957): Research student at Cambridge University, later Professor of

Physics at Oxford University, knighted in 1941.

VAN DEN BROEK, ANTONIUS JOHANNES (1870–1926): Private teacher of physics in Noordwyk, Holland.

WELSBACH, CARL AUER VON, (1858–1929): Austrian chemist, inventor of gas-mantle lamp.

WILSON, HAROLD ALBERT (1874–   ): Research student in physics at Cambridge University, later Professor of Physics at King's College, London, McGill University, Montreal, the University of Glasgow, and the Rice Institute, Houston, Texas.

WINKLER, CLEMENS (1838–1904): Professor of Chemistry at the School of Mines in Freiberg, Germany.

ORIGIN OF NAMES AND SYMBOLS*

Ac   actinium, 1900; from the Greek *aktis,* ray.

Al   aluminum, 1825; from the Latin *alumen,* substance having an astringent taste.

Am   americium, 1944; named for America where discovered.

Sb   antimony, 15th century; from the Greek *antimonos,* opposed to solitude, as antimony is always associated with other minerals.

A   argon, 1894; from the Greek *argos,* neutral or inactive.

As   arsenic, 13th century; from the Greek *arsenikon,* valiant or bold, from its action on other metals.

At   astatine, 1940; from the Greek *astatos,* unstable.

Ba   barium, 1808; from the Greek *barys,* heavy.

Bk   berkelium, 1949; named for Berkeley, California.

Be   beryllium, 1797; named for the mineral beryl.

Bi   bismuth, 15th century; from the German *weisse Masse,* white mass; called by miners *wismut;* changed to *bismat* when latinized.

B   boron, 1808; from the Arabic *bawraq* or Persian *burah,* white.

Br   bromine, 1826; from the Greek *bromos,* a stench.

* Dates are those of isolation of the element when known; otherwise of discovery.

Cd   cadmium, 1817; from the Latin *cadmia*, calamine, a zinc ore.

Ca   calcium, 1808; from the Latin *calcis*, lime.

Cf   californium, 1950; named for the State and the University of California.

C   carbon, prehistoric; from the Latin *carbo*, coal.

Ce   cerium, 1804; named from the asteroid, Ceres, discovered in 1801.

Cs   cesium, 1860; from the Latin *caesius*, sky blue.

Cl   chlorine, 1808; from the Greek *chloros*, grass green.

Cr   chromium, 1797; from the Greek *chroma*, color.

Co   cobalt, 1735; from the Greek *kobolos*, a goblin.

Cu   copper, prehistoric; from the Latin *cuprum*, copper.

Cm   curium, 1944; named for Marie and Pierre Curie.

Dy   dysprosium, 1886; from the Greek *dysprositos*, hard to get at.

E   einsteinium, 1952; named for Albert Einstein.

Er   erbium, 1843; named for Ytterby, town in Sweden.

Eu   europium, 1900; named for Europe.

Fm   fermium, 1953; named for Enrico Fermi.

F   fluorine, 1886; from the Latin *fluere*, to flow.

Fr   francium, 1939; named for France, native country of its discoverer.

Gd   gadolinium, 1886; named for J. Gadolin, Finnish chemist.

Ga   gallium, 1875; named for Gaul or France.

Ge   germanium, 1886; named for Germany.

Au   gold, prehistoric; from the Anglo-Saxon *gold;* symbol from Latin *aurum.*

Hf   hafnium, 1922; from *Hafnia,* Latin for Copenhagen.

He   helium, 1895; from the Greek *helios,* the sun.

Ho   holmium, 1879; from *Holmia,* latinized form of Stockholm, near which many rare earth minerals are found.

H   hydrogen, 1766; from the Greek *hydro genes,* water former.

In   indium, 1863; named from its indigo-blue spectrum line.

I   iodine, 1811; from the Greek *iodes,* violet-like.

Ir   iridium, 1804; from the Latin *iridis,* rainbow; named for the iridescent color of its salts.

Fe   iron, prehistoric; from the Anglo-Saxon *iren* or *isen;* symbol from the Latin *ferrum.*

Kr   krypton, 1898; from the Greek *kryptos,* hidden.

La   lanthanum, 1839; from the Greek *lanthanein,* to be concealed.

Pb   lead, prehistoric; from the middle English *led;* symbol from the Latin name *plumbum.*

Li lithium, 1817; from the Greek *lithos,* stone.

Lu lutetium, 1905; from *Lutetia,* ancient name of Paris.

Mg magnesium, 1808; from the Latin *Magnesia,* a district in Asia Minor.

Mn manganese, 1774; from the Latin *magnes,* magnet.

Mv mendelevium, 1955; named for Dmitri Mendeleyev who first devised the Periodic Table.

Hg mercury, prehistoric; from the Latin *Mercurius,* the god and planet.

Mo molybdenum, 1782; from the Greek *molybdos,* lead.

Nd neodymium, 1885; from the Greek *neos,* new, and *didymos,* twin.

Ne neon, 1898; from the Greek *neos,* new.

Np neptunium, 1940; named for the planet Neptune.

Ni nickel, 1750; from the German *Nickel,* a goblin or devil.

Nb niobium, 1801; named for Niobe, daughter of Tantalus, because it is always found associated with tantalum.

N nitrogen, 1772; from the Latin *nitro,* native soda, and *gen,* born.

No nobelium, 1957; named for Alfred Nobel.

Os osmium, 1804; from the Greek *osme,* a smell, from the odor of its volatile tetroxide.

O oxygen, 1774; from the Greek *oxys,* sharp, and *gen,* born.

Pd palladium, 1803; named for the planetoid Pallas, discovered in 1801.

P   phosphorus, 1669; from the Greek *phosphoros,* light bringer.

Pt   platinum, 1735; from the Spanish *plata,* silver.

Pu   plutonium, 1940; named for Pluto, the second trans-Uranus planet, since it is the second trans-uranium element.

Po   polonium, 1898; named for Poland, the native country of its discoverer.

K   potassium, 1807; from the English *potash;* symbol from the Latin *kalium.*

Pr   praseodymium, 1885; from the Greek *praseos,* leek green, and *didymos,* a twin.

Pm   promethium, 1947; named for Prometheus, the fire bringer of Greek mythology.

Pa   protactinium, 1917; from the Greek *protos,* first, and *actinium,* because it disintegrates into actinium.

Ra   radium, 1898; from the Latin *radius,* ray.

Rn   radon, 1900; so named because it comes from radium.

Re   rhenium, 1924; named from the Rhine province belonging to Germany (*Rhenus* in Latin).

Rh   rhodium, 1804; from the Greek, *rhodon,* a rose, from the color of the aqueous solutions of its salts.

Rb   rubidium, 1860; from the Latin *rubidus,* red.

Ru   ruthenium, 1845; named for Russia (*Ruthenia* in Latin).

Sm   samarium, 1879; named for Samarski, a Russian engineer.

Sc   scandium, 1879; named for the Scandinavian peninsula by its Swedish discoverer.

Se   selenium, 1817; from the Greek, *selene,* the moon.

Si   silicon, 1823; from the Latin *silex,* flint.

Ag   silver, prehistoric; from the Anglo-Saxon *seolfor;* symbol from the Latin name *argentum.*

Na   sodium, 1807; from the medieval Latin *soda;* symbol from the Latin *natrium.*

Sr   strontium, 1808; named for the town of Strontian, Scotland.

S   sulfur, prehistoric; from the Latin *sulphur.*

Ta   tantalum, 1802; named for Tantalus of Greek mythology because the metal was hard to isolate, therefore tantalizing.

Tc   technetium, 1937; from the Greek *technetos,* artificial, because this was the first element to be made artificially.

Te   tellurium, 1782; from the Latin *tellus,* the earth.

Tb   terbium, 1843; named for Ytterby, town in Sweden.

Tl   thallium, 1862; from the Greek *thallos,* a young shoot.

Th   thorium, 1819; from Scandinavian mythology, *Thor.*

Tm   thulium, 1879; from the Latin *Thule,* the most northerly part of the habitable world.

Sn   tin, prehistoric; actual origin of name unknown; symbol from the Latin *stannum*.

Ti   titanium, 1791; from Greek mythology, *Titans,* the first sons of the earth.

W   tungsten or wolfram, 1783; from the Swedish *tung sten,* heavy stone; from the mineral wolframite.

U   uranium, 1789; named for the planet Uranus.

V   vanadium, 1830; named for the goddess *Vanadis* of Scandinavian mythology.

Xe   xenon, 1898; from the Greek *xenos,* strange.

Yb   ytterbium, 1905; named for Ytterby, town in Sweden.

Y   yttrium, 1843; named for Ytterby, town in Sweden.

Zn   zinc, prehistoric; from the German *Zink,* akin to *Zinn,* tin.

Zr   zirconium, 1824; from the Arabian *Zerk,* a precious stone.

SUGGESTED READINGS

The Neutron Story, Donald J. Hughes (Science Study Series, Doubleday Anchor Books, 1959)
>   Deals with some of the nuclear physics which developed in the period after the end of this book.

Explaining the Atom, Selig Hecht (Viking Press, 1948, revised edition 1954)
>   Another good account of more recent nuclear physics.

How Old Is the Earth?, Patrick M. Hurley (Science Study Series, Doubleday Anchor Books, 1959)
>   Isotopes and radioactivity as used to study the history of the earth.

The Interpretation of Radium and the Structure of the Atom, Frederick Soddy (4th Edition, G. P. Putnam's Sons, 1922)
>   Based on a series of lectures given at the University of Glasgow in 1908, and brought up to date to 1920. Likely to be found in libraries.

Radioactivity: The Transformation of Elements and *Radiochemistry and the Discovery of Isotopes,* Alfred Romer, editor (Classics of Science Series, Dover Publications, in preparation)
>   Reprinting the original reports of some of the most important investigations described in this book.

INDEX

SCIENCE STUDY SERIES